Steve:
Compliments of
the author
Dick

HANDBOOK OF INDUSTRIAL NOISE MANAGEMENT

by

Richard K. Miller

Richard K. Miller & Associates, Inc.
Consultants in Acoustics
Atlanta, Georgia

Edited and Preface by
Albert Thumann, P.E.
Bechtel Corporation
Louisville, Kentucky

Introduction by
Truett Smith, Attorney-At-Law
Grant, Smith & Shiver
Elberton, Georgia

THE FAIRMONT PRESS
134 Peachtree Street, NE
Suite 822
Atlanta, Georgia 30303

HANDBOOK OF INDUSTRIAL NOISE MANAGEMENT

ISBN: 0-915586-02-9
LIBRARY OF CONGRESS CATALOG CARD NO.: 76-6804

This book is dedicated to
my wonderful parents.

PREFACE

The industry-wide total cost to achieve compliance with the OSHA 90 dBA regulation has been estimated to be in excess of $13 billion, and the industry-wide liability for worker's compensation claims due to noise-induced hearing loss is expected to exceed this dollar value. Industry has had few guidelines to the legal and cost/benefit approaches to noise control. *Handbook of Industrial Noise Management* by Richard K. Miller fills this void. It is the first book to address such problems as defining when a noise control is not feasible, and assessing the costs and benefits of noise control.

The author has been called on as an "expert witness" in many OSHA-related litigations. This in-depth, first-hand knowledge is applied in his handbook. A library of noise control essentials should contain books on the technical, management and medical aspects of noise control. *Handbook of Industrial Noise Management* is a necessary addition to this library.

June 1976 ALBERT THUMANN, P.E.

CONTENTS

INTRODUCTION

The problem which American industry faces in complying with noise regulations is neither a straightforward engineering problem nor a straightforward legal problem. Hence, the industrial executive is called upon to prudently channel the expertise of legal and engineering professionals, and to coordinate noise compliance efforts, compromising neither the best interests of the company nor the well-being of employees. This is certainly a challenging responsibility. Indeed one might reasonably ask whether the manager who establishes the goals of treating his employees well, complying with the law, and at the same time making money, is not facing an impossible dilemma. This is a question which Richard Miller has addressed for many years. The management guidelines presented in the chapters which follow have evolved from his experience as a consultant to many of the largest industries in the United States, both prior to and after the promulgation of the Occupational Safety and Health Act of 1970. He has achieved recognition as an expert in the area of industrial noise control by virtue of his participation in the field, and has acquired a broad working knowledge of OSHA enforcement and legal procedures through preparation of compliance plans, and by serving as an expert witness for both industry and the U.S. Department of Labor.

I recently had the opportunity of working with Rich Miller in an OSAHRC trial which saw the involvement of five OSHA expert witnesses, and compliance issues related to both engineering feasibility and measurement accuracy. Rich did not rely on his credentials to support his professional opinion and testimony, but obviously did his homework for the trial and presented strong technical support for our mutual client. This thoroughness is reflected in this handbook, in which he goes beyond the technical aspects of noise control in a very readable, interesting, and well-organized format. The book represents perhaps the first attempt to deal with all aspects of an industrial noise program by one individual. I personally don't know of any other publication which addresses the issues of noise control feasibility, OSHA compliance procedures, and protection against unjust worker's compensation claims for hearing loss in such a manner. I feel that this handbook will be equally valuable for industrial executives, engineers, and attorneys as an aid in understanding their respective noise control responsibilities and in guiding their participation in noise abatement programs.

TRUETT SMITH
September 1976 Attorney-at-Law

Chapter 1
THE NEED FOR NOISE MANAGEMENT

The Occupational Safety and Health Act (OSHA) may be the most controversial federal law in history. Some industrial managers feel OSHA provides them with two alternatives; either they can deliberately go bankrupt trying to retrofit old equipment and buying new equipment to comply with the standard, or they can wait for the OSHA compliance officer to shut their plant down. The controversy generated by OSHA has provided prime editorial material for the nation's leading magazines and hot campaign issues for election campaign platforms. Even President Ford addressed an audience by saying, "I'm sure you'd all like to see OSHA dumped in the ocean." Most controversial of the OSHA regulations is the noise control standard, over which controversy raged for over four years with regard to revising the original 90 dBA* noise limits, and for which estimated compliance costs to industry range up to 35 million dollars.

Nobody disagrees that maintaining a safe and healthy work environment is both worthwhile and necessary. Many businessmen complain about the manner in which OSHA has implemented the regulations, and attacks are focused on isolated errors such as the case of a small grocery store operator who was forced to install a

---

* 90 dBA is the limit for an 8 hour work day. Higher sound levels are allowable for lower exposure time (see Chapter 2, Table 2.1).

separate bathroom for his only employee, his wife. While many complaints may have a legitimate basis, primary opposition to the OSHA standards is motivated by fear related to economics. Indeed, the economic statistics are astounding.

The economic impact study recently published by OSHA concluded that industry would have to spend $10.5 billion to comply with engineering requirements for a 90 dBA standard, and $18.5 billion to meet an 85 dBA standard. Rigid enforcement of such a standard would have a tremendous impact on industry's safety and health budget, which is estimated to be $3.2 billion for 1976, and represents approximately 3% of the total capital investment for manufacturing firms. Enforcement of the noise standard shows an increasing trend, with 43% of the 10,641 noise samples between January 1973 and March 1975 resulting in proposed penalties of $169,042.

While impressive, these economic statistics related to order-of-magnitude problems faced by the nation's business and government leaders are only of academic interest to the plant manager and safety engineer. An industrial manager must be profit oriented, and his attention is focused on the thousands of dollars which *his* plant must spend, not the billions which will be spent nationwide. The safety engineer must organize his plant's noise abatement program for *optimum* results; maximum protection of employee hearing with legal compliance, for minimum expenditure. The effectiveness of a hearing conservation and noise control program is often not related to either the dollar commitment involved or state-of-the-art technology. How many plants have undertaken costly long term noise control engineering programs, but have incurred tremendous potential worker's compensation claims by ignoring the economical interim measures of hearing protective devices and audiometric testing? How often are costly noise control measures implemented on a "trial and error" basis and fail, when positive results could have been achieved with proper training of in-house personnel or by hiring a qualified acoustical consultant? It is astonishing to realize how many plants jump into an acoustical engineering program without first establishing noise abatement priorities or performing an economic analysis. Recent advances in the state-of-the-art in noise control engineering, largely brought about by the positive influence of OSHA, have provided solutions for a majority of industrial noise problems. What is required now is proper implementation of the state-of-the-art technology and effective *noise management*.

CHANGING STANDARDS: OSHA OVERVIEW

Federal involvement in occupational noise regulation had its beginning in May 1969, with the passage of the Walsh-Healey Public Contracts Act[1]. Then in December 1970, the Occupational Safety and Health Act[2] (OSHA) incorporated verbatim into law the noise regulations and standards of the Walsh-Healey Act. (The reader may find a copy of the noise section of this regulation in Appendix A.) Since its establishment, ambiguities and controversies have surrounded the noise control criteria of OSHA. Many specific questions have arisen, such as:

1. Does the 90 dBA noise exposure criteria really protect workers from hearing loss?

2. Why aren't hearing protective devices an acceptable long range occupational noise exposure solution?

3. What is the definition of technical and/or economic feasibility?

4. How much time is to be allowed for achieving compliance with the law?

On August 14, 1972, the Department of Health, Education and Welfare presented to the Department of Labor a noise criteria document prepared by the National Institute for Occupational Safety and Health[3] (NIOSH). The NOISH criteria document recommended lowering the 90 dBA limit for eight hour exposure to 85 dBA. The NIOSH criteria recognized that continuous exposure to noise sources at levels above 85 dBA may be harmful to the hearing health of American workers.

In response, the Department of Labor named a Noise Advisory Committee composed of four labor representatives, four employee representatives, two state and two federal governmental representatives, and three professionals. In December 1973, after spending more than a full year reviewing the NIOSH criteria and holding hearings, a report was issued which maintained the 90 dBA limit for eight hour exposure. The representatives of labor and the Air Force expressed objection to the report's conclusions.

During this same period, a report[4] was being finalized for the Department of Labor by Bolt, Beranek and Newman, Inc., a Cambridge, Massachusetts consulting firm. This report, published in January 1974, indicated that the cost for complying with the OSHA law for 19 industrial classifications could be expected to total $31 billion for an 85 dBA standard, as compared to $13.4 billion for the existing 90 dBA standard. In July 1976 an economic impact statement by the same firm revised these cost estimates to $18.5 and $10.5 billion respectively.

Concurrently, the Environmental Protection Agency (EPA) involvement with noise was gaining momentum. EPA, under the Noise Control Act of 1972, was given the authority and direction to "promote an environment for all Americans" which would be "free from noise that might jeopardize their health and welfare." The act gave to EPA the responsibility for coordinating all federal noise control programs. In March 1974, EPA released, "Information on Levels of Environmental Noise Requisite to Protect Public Health and Welfare With an Adequate Margin of Safety."[5] This document also proposed noise levels much lower than those established by OSHA.

The first change OSHA proposed to the original Walsh-Healey Act came on October 24, 1974. OSHA published a new proposed standard[6] which maintained 90 dBA as the permissible noise exposure level for an eight hour day, but reduced the allowable exposure for a 16 hour period to 85 dBA. Appendix B presents this revised noise criteria. As indicated, the criteria retained the 5 dB doubling rate for exposure times, i.e.:

95 dBA - 4 hours
90 dBA - 8 hours
85 dBA - 16 hours

A formula for computing the exposure time for levels not covered in the original table was also given. In addition, the criteria designated specific limitations for impulse noise. Specifically, it limited exposure to impulses of 140 dBA to 100 per day and permitted a ten-fold increase in the number of impulses for each 10 dB decrease in the peak sound pressure level of the impulse (e.g., the number of impulses allowed at 120 dB would be 10,000 per day).

In the revised noise criteria, OSHA addressed the controversy surrounding the use of hearing protectors in lieu of engineering noise controls by stating simply that the use of hearing protectors as a primary means of controlling noise exposure is not good industrial practice. OSHA expanded the original standard by proposing that when it is not feasible to reduce noise exposure to within the designated limits through engineering and/or administrative controls, such controls must be utilized to the maximum extent which *is* feasible and must be supplemented by personal protective equipment.

Another significant aspect of the new criteria was to require audiometric testing for all employees with an eight hour noise exposure of 85 dBA or higher and for all employees who wear personal protective equipment. The rationale for the audiometric testing program is that it will detect changes in the hearing level of workers, thus enabling corrective action as well as providing data on the effectiveness of the noise criteria.

The criteria also contains regulations on monitoring and record keeping.

WORKER'S COMPENSATION

Another event is occurring which may have a far greater impact than OSHA: the passage and enforcement of worker's compensation laws. One plant has already experienced claims filed for a dollar value exceeding one million dollars. With 75% of the nation's workforce exposed to sound levels above 85 dBA, potential worker's compensation liability for hearing loss has been estimated to be on the order of $12 billion.

Worker's compensation laws in most states now make a hearing loss compensable. The cost to industry for potential liability will underscore the cost to comply to a more stringent noise standard. As an example, in the District of Columbia, compensation allowable for the loss in both ears was reported to range up to $33,000![7]

The interesting point about the worker's compensation law is that if the employer does not establish that his employee had a hearing loss *prior* to employment, the employer is liable for the loss. This means that audiometric testing will be required for all newly hired workers, not necessarily because of the law, but in order to establish liability. The economic incentive will, no doubt, cause employers to be more concerned with enforcing proper use of ear protectors as well as lowering the employees' exposure to noise.

WHAT'S AHEAD?

Noise control regulations are certainly going to continue to change. New data which is made available through audiometric testing programs, and new quieter products which are becoming available, will provide the backup needed to allow the promulgation of more stringent noise criteria. Many trade-offs among energy, economic and environmental considerations will no doubt take place.

The controversies of the past several years have served to focus public attention on the noise control issue. The public awareness and the heated discussion among government agencies have, in fact, fueled an interest in noise which was previously not thought to be attainable. This "healthy conflict" should lead to the designing of noise standards which will reflect a regard for economic impact as well as safety and technology.

REFERENCES

1. Walsh-Healey Public Contracts Act, *Federal Register* Vol. 34, No. 96, May 20, 1969.

2. Occupational Safety and Health Act, Williams-Steiger Occupational Safety and Health Act of 1970, PL 91-596, December 29, 1970.

3. "Occupational Exposure to Noise," NIOSH Criteria Document, Department of Health, Education and Welfare, HSM 73-11001, August 14, 1972.

4. "Impact of Noise Control at the Workplace," BBN Report No. 2671, available from OSHA, January 1974.

5. "Information on Levels of Environmental Noise Requisite to Protect Public Health and Welfare With an Adequate Margin of Safety," Environmental Protection Agency Publication 550/9-74-004, March 1974.

6. "Occupational Noise Exposure - Proposed Requirements and Procedures," *Federal Register*, Vol. 39, No. 207, October 24, 1974.

7. "Noise-Induced Hearing Loss - Role of the Physician in Workmen's Compensation Cases," American Medical Association's Sixth Congress on Environmental Health, April 1969.

8. Thumann, A., "A Review of Currently Changing Standards," delivered at the International Pollution Engineering Exposition and Congress, October 22, 1975.

Chapter 2
OSHA COMPLIANCE STRATEGIES

The Williams-Steiger Occupational Safety and Health Act of 1970 (Public Law 91-596) was established "to assure safe and healthful working conditions for working men and women." The Occupational Safety and Health Administration (OSHA) of the U.S. Department of Labor was delegated the responsibility of implementing and enforcing the law.

Title 29 CFR, Section 1910.95 of this original law (refer to Appendix A) pertains to the protection of workers from potentially hazardous occupational noise. The revised OSHA noise guidelines (reprinted in Appendix B), which serve to clarify some of the vague or unclear portions of the original OSHA law, are considered here.

PERMISSIBLE EXPOSURE LIMITS

While the OSHA regulations establish a maximum noise level of 90 dBA for a continuous 8 hour exposure during a working day, higher sound levels are allowed for shorter exposure times. For example, in the manufacture of concrete pipe, employees are seldom exposed to continuous noise levels. Maximum noise exposure generally occurs only during the duty cycle of a machine. Thus, it is necessary to compute the employee's noise dose, or percent allowable exposure for actual operation.

A noise dose, D, is defined as the actual exposure time

TABLE 2.1

PERMISSIBLE NOISE EXPOSURE TIMES

Sound Level (dBA)	Time (hrs)
85	16.00
86	13.93
87	12.13
88	10.57
89	9.18
90	8.00
91	6.97
92	6.07
93	5.28
94	4.60
95	4.00
96	3.48
97	3.03
98	2.83
99	2.25
100	2.00
101	1.73
102	1.52
103	1.32
104	1.15
105	1.00
106	0.87
107	0.77
108	0.67
109	0.57
110	0.50
111	0.43
112	0.38
113	0.33
114	0.28
115	0.25

divided by the allowable exposure time. The allowable exposure time, from Table G-16 of the OSHA regulation, is given in Table 2.1. The employee noise exposure exceeds the OSHA limits if the noise dose exceeds unity (1), or 100%.

$$D = \frac{C}{T} \tag{2-1}$$

where: D = noise dose
C = actual duration of exposure in hours
T = noise exposure limit in hours

Where the daily exposure is due to two or more noise levels, the ratio for each level is added to compute the total noise dose:

$$D = \frac{C_1}{T_1} + \frac{C_2}{T_2} + \cdots \frac{C_n}{T_n} \tag{2-2}$$

EXAMPLE 1

Calculate the daily noise dose for the operator of a machine which produces 160 parts during an 8 hour period. The machine sound level is 95 dBA, and the noise-producing cycle is 2 minutes in duration per part.

Solution:

The total exposure time is:

$$T = 160 \text{ parts} \times 2 \text{ minutes per part}$$

$$= 320 \text{ minutes}$$

$$= 5.33 \text{ hours}$$

The allowable exposure time, from Table 2.1, is 4

hours. The noise dose, from equation (2-1) is:

$$D = 5.33 \div 4$$

$$D = 1.33$$

Thus, the operator noise exposure exceeds the OSHA limits.

EXAMPLE 2

During a single cycle, a machine generates 102 dBA for 6 minutes (.1 hour) and 106 dBA for 3/4 minute (.0125 hour). The operator may operate the machine for 1 to 8 hours per day. The machine has four cycles per hour. Compute the operator's noise dose.

Solution:

From Table 2.1, the allowable exposure time for 102 dBA is 1.52 hours, and is .87 hour for 106 dBA. Applying equation (2-2), his noise dose each hour is:

$$D = 4\left(\frac{.1}{1.52} + \frac{.0125}{.87}\right)$$

$$D = .32$$

His daily exposure is equal to his hourly exposure times the number of hours the machine is operated. A summary of his total exposure time for various daily operations is as follows:

Machine Operation (Hours)	*Daily Noise Dose*
1	32%
2	64%
3	96%
4	128%

5	160%
6	192%
7	224%
8	256%

Thus, if the daily operation exceeds 3.1 hours, it would not be in compliance with OSHA.

It should be pointed out that if the employee is engaged in another noisy operation during part of his work day, the noise dose of that operation must also be added to the above value to obtain the total noise dose.

ADMINISTRATIVE CONTROLS

The OSHA regulations require the implementation of either engineering or administrative controls where feasible. Administrative controls are those which involve changes in employee or production operation schedules which would reduce the employee noise dose. The following are examples:

1. Arrange work schedules so that employees working the major portion of a day at or very close to the 90 dBA limit are not exposed to higher noise levels.

2. Insure that employees who have reached the upper limit of duration for a high noise level in accordance with Table G-16 of the OSHA regulations work the remainder of the day in an environment with a noise level well below 90 dBA.

3. Where the man-hours required for a job exceed the permissible time for one man in one day for the existing sound level, divide the work among two, three, or as many workers as are needed, either

successively or together, to keep individual noise exposure within permissible time limits.

4. If less than fulltime production of a noisy machine is needed, arrange to run it a portion of each day rather than all day for a part of the week.

5. Perform occasional high noise level producing operations at night or at other times when a minimum number of employees will be exposed.

Measures such as these can often be instituted at little cost or effort, simply by introducing noise exposure as a factor in production planning. While not as satisfactory as controlling noise at its sources, administrative control measures are more easily enforced than is the requirement to wear personal protective equipment. For this reason they are preferred.

The revised OSHA regulations reduce the potential for administrative controls in many cases. In the original regulation, unlimited exposure was allowed for all sound levels below 90 dBA. In the revised standard, exposure limits are established for lower sound levels.

As an example, assume two employees who work in different areas are exposed to sound levels of 93 dBA and 88 dBA respectively. Thus, the first worker is exposed to excessive noise, while the second worker is not. As an administrative control, it is proposed that the two employees rotate their tasks, or split their time such that each works in each area for four hours during the work day. Under the original regulation, the noise dose of each employee would be (equation 2-1):

$$D = \frac{\text{time exposed}}{\text{allowable time}} = \frac{4}{6} = 0.66$$

and compliance would be acheived.

With the revised regulation, Table 2.1:

$$D = \frac{4}{6} + \frac{4}{10.57} = 1.04$$

Thus, the noise exposure of both employees exceeds the revised OSHA regulation.

Since risk of hearing loss does not disappear for noise exposure below 90 dBA, the proposed revision is a more realistic measure for assessing daily exposure. Hence, administrative controls are generally effective only where a significant number of the employees of a plant are exposed to sound levels below 85 dBA.

Several practical problems may eliminate the possibility of utilizing administrative controls as a noise compliance option:

1. *Labor Agreements*. Administrative controls may be in conflict with employees' bargained for rights and benefits under existing labor agreements. Necessary movement of employees may violate bargained for conditions of employment affecting rates of pay, promotion, demotion and lay-off, hours of work, job jurisdiction, union representation, and overtime provisions, as well as other entitlements.

2. *Safety*. Teamwork of skilled employees within a production process may be disrupted. Such groups of employees may rely upon one another's knowledge of equipment operation, communication methods and procedures, and proper safety precautions.

3. *Lack of Job Continuity*. In a continuous operation, the time lapse, or period when a job or jobs would be unmanned, could seriously affect the safety, efficiency and economy of the operation.

4. *Training*. Employee training is a continuous proc-

ess. Each day served on a job is another step in becoming a more skilled craftsman. Movement between process areas may affect the acquisition of these skills.

VARIANCES

Obtaining a variance, or authorization to proceed with safety procedures or conditions other than those dictated by the regulations, provides another approach to dealing with the OSHA standards. Variances are only applicable from the date of issuance, and cannot be utilized to dismiss existing citations. It is necessary that variances be granted prior to an OSHA inspection or issuance of a citation.

There are four types of variances. These are *temporary* variances, *permanent* variances, *experimental* variances, and *national defense* variances. The details of each of these types of variances are as follows:

TEMPORARY VARIANCE

1. The employer must establish that he is unable to comply with a standard by its effective date due to one or more of the following:

    a. Unavailability of professional personnel.

    b. Unavailability of technical personnel.

    c. Unavailability of needed materials.

    d. Unavailability of equipment.

    e. Required construction or alteration of facilities which cannot be completed before the effective date of the standard.

2. The employer must show that he is taking all available steps to protect his employees from the hazards covered by the standard.

3. No temporary variance may be in effect for longer than the time needed to comply with the requirement, or for 1 year (whichever is shorter). The variance may be renewed twice.

PERMANENT VARIANCE

When equivalent or better protection is provided than would be obtained by compliance with the standard, a permanent variance may be obtained.

EXPERIMENTAL VARIANCE

The Secretary of Labor may grant a variance whenever he determines - or the Secretary of Health, Education, and Welfare certifies - that a variance is necessary to permit an employer to participate in an approved experiment. The purpose of the experiment must be to demonstrate or validate new and improved techniques to protect the safety or health of employees.

NATIONAL DEFENSE VARIANCE

The Secretary of Labor may allow reasonable variations, tolerances, and exemptions to and from any or all provisions of the OSHA regulations to avoid serious impairment of the national defense. This type of action cannot be in effect for more than 6 months without notification of affected employees.

In dealing with noise problems, application for a temporary variance may be advisable where implementation of administrative or engineering measures for compliance is anticipated within a 1-3 year period. It must be recognized, however, that a strong possibility of a citation exists if the problem is not abated when the variance expires.

It is generally not advisable to apply for a variance where engineering noise reduction methods are not feasible. A temporary variance would require abatement within a maximum of three years, which would not be possible if there exist no feasible solutions. A permanent variance would not be issued solely on the basis that a feasible solution does not exist. It must also be shown that equivalent or better protection is provided than would be obtained if the noise were abated.

While there is no provision at the federal level for obtaining a permanent variance based on non-feasibility, other than by extension of one year variances, some states which have OSHA enforcement authorities have provided for permanent variances. Such variances have been obtained by industry for operations such as can container manufacturing, sand blasting, and arc-air cutting.

ENFORCEMENT BY OSHA

Enforcement of the Occupational Safety and Health Act is maintained by means of inspections by an OSHA compliance officer. When his inspection reveals conditions which he believes do not comply with the Act, a citation will be issued.

A representative of management should be appointed to provide information required by the inspector and to accompany him on the plant inspection tour. He may have an employee representative accompany him also. Presentation by the inspector of proper identification should always be expected. Cases of fraudulent inspections by individuals not affiliated with OSHA have been reported. OSHA is also required to keep confidential any information designated as containing trade secrets.

The citation will contain three important items:

1. The alleged violation(s).

2. The date by which the alleged violation(s) must be abated.

3. A proposed penalty.

It the employer recognizes violation of the Act and intends to comply with the citation stipulations, he may proceed directly with an abatement program. Generally, the employer will be requested to submit a noise abatement compliance plan for approval by OSHA to assure that his approach to compliance is reasonable.

COMPLIANCE PROGRAM

The compliance program is a response by an industry to a citation which states that they intend to comply with the Act. The program provides a commitment to the engineering or administrative control measures required for compliance, and related schedules and expenditures.

Generally, the development of a compliance program involves:

1. An engineering feasibility study performed by a consultant or in-house engineers to:

    a. Establish state-of-the-art engineering noise

reduction techniques for each cited operation.

b. Develop conceptual designs for noise abatement of each cited operation.

c. Document technical justification where noise reduction is not feasible. (The establishment of engineering feasibility is discussed in Chapter 3.)

2. The development of conceptual engineering designs for the noise reduction of each cited operation. Specific designs may be developed after the settlement agreement is reached.

3. A hearing conservation program must be developed. As discussed in Chapters 4 and 5, this involves:

   a. Audiometric testing.

   b. Hearing protective devices.

4. The industry must make a commitment to:

   a. The implementation dates of the hearing conservation, engineering and administrative programs.

   b. The costs associated with the program.

The compliance program document should include the following items:

1. *Introduction*. The circumstances regarding the OSHA citation and noise abatement program should be reviewed.

2. *Noise Abatement Policy*. The general policies pertaining to noise control and hearing conservation should be outlined.

3. *Engineering Controls.* The conceptual approaches to be implemented to achieve noise reduction for each of the cited operations should be presented.

4. *Administrative Controls.* The proposed use of administrative controls to reduce employee noise exposure should be outlined. Where operational requirements or labor union contracts preclude the use of administrative controls, these items should be discussed.

5. *Feasibility.* The results of the engineering feasibility study should be presented for all operations where engineering or administrative controls are not considered feasible.

6. *Hearing Conservation.* The program for the use of personal hearing protective devices and audiometric testing should be presented.

7. *Previous Efforts.* Past noise control and hearing conservation efforts should be described in support of the good faith effort of the employer.

8. *Organization.* The company's administrative and engineering personnel responsible for the attainment of compliance should be presented.

Regional and area OSHA requirements on the specific content of the compliance program may vary. In most cases, a well-defined, but general outline of abatement items is sufficient. A detailed model compliance program is presented in Appendix C.

FILING A NOTICE OF CONTEST

The employer has the legal privilege of contesting the citation, and will be advised of the following upon receipt of a citation:

> You have the right to contest any or all parts of either the citation(s) or the proposed penalty(s) before the Occupational Safety and Health Review Commission. The Review Commission is an independent agency with authority to issue decisions regarding citations and proposed penalties. If you do contest, you should submit a letter to the Area Director within 15 working days after receipt of the certified mail notice regarding proposed penalty(s). If you fail to contest within the 15 working day period, the citation(s) and the penalty(s) as proposed, shall be deemed to be a final order of the Review Commission and not subject to review by any court or agency.

The Occupational Safety and Health Review Commission (OSAHRC) is an independent governmental administrative agency consisting of three commissioners in Washington, D.C. and assisted by approximately 50 Federal Law Judges throughout the country.

The reasons for filing a notice of contest for a noise citation may be summarized as follows:

A. Engineering or administrative measures are not feasible to reduce the employee noise exposure, based on:

   1. Engineering and technical considerations.

   2. Production requirements.

   3. Economic analysis.

   It has been the position of the Secretary of Labor that engineering controls must be implemented if (significant) noise reduction can be achieved, whether or not exposure is reduced to within the allowable values of Table G-16. A recent court ruling established that such controls were not necessary if the allowable exposure would not be achieved. However, this ruling is subject to OSAHRC review. (*Note:* Engineering feasibility is discussed in Chapter 3.

B. The operations observed during the inspection were not characteristic of the normal plant operations due to plant start-up or shut-down, unusual maintenance conditions, emergency conditions, production irregularity, etc. While a safety hazard would not be allowable for a short duration, health risk due to noise exposure is generally present only with prolonged exposure, and a short duration exposure would not ordinarily be hazardous.

C. The occupation was outside the jurisdictional domain of the U.S. Department of Labor:

1. Governmental employees.

2. Transportation (regulated by the U.S. Department of Transportation

3. Mining (regulated by the Coal Mine Health and Safety Act, and the Federal Metal and Nonmetallic Safety Act).

D. An active noise abatement program was in effect, hearing protective devices were being utilized to provide interim protection until engineering and/or administrative controls were perfected. It would be necessary to show reasonable progress of the program, and to provide supportive written documentation.

E. The proposed penalty was excessive:

1. It did not reflect hte severity of the health hazard.

2. A good faith effort had been shown.

F. The compliance date for correction of the alleged violation was unreasonable.

G. A technically incorrect procedure was followed, such as incorrect information within the citation (dates, referenced sections of the Code of Federal Regulations, etc.).

H. The noise dose of the cited employee did not exceed the values of Table G-16 because the compliance officer's analysis was incorrect. There are several reasons why the officer's analysis might be incorrect, including:

1. No sound level measurements were performed.

2. The instrument was not calibrated in the field prior to and subsequent to the measurements.

3. The instrument was not calibrated in accordance with the procedures of the Bureau of Standards.

4. The sound level measurement was performed at a location which is not representative of where the employee is stationed, or is not a true representation of the zone of hearing of the employee.

5. Environmental conditions adversely influenced the measurement: excessive air velocities at the microphone (approximately 3-8 feet per second without windscreen and 15-25 feet per second with windscreen); excessive vibration

or impact of instrument; excessive moisture, temperature extremes, etc.

6. The accuracy of the instrument was not adequate for the sound levels present.

7. The sound levels fluctuate with time, and the noise dose was not correctly obtained. A visual average of the sound level meter fluctuations is generally not adequate.

8. The noise dose was not correctly calculated.

9. A dosimeter was not used to establish the noise dose.

10. A dosimeter was used, and it cannot be ascertained that recorded values were due to the sound level in question rather than employee tampering, impact of the microphone, voices, etc.

11. Dosimeter measurements were not verified by sound level meter readings and time measurements.

12. The dosimeter readings were inaccurate due to proximity to the noise source, or body reflections.

Note: Generally, challenges of instrument accuracy are valid only where the ascertained noise exposure marginally exceeds the standard. A calibrated instrument would not be suspect of errors of 10-15 dB due to tolerances.

If a notice of contest is filed to the Department of Labor, it is submitted to OSAHRC. Within 20 days after receiving the notice of contest, the Secretary of Labor must file a written complaint with OSAHRC. The complaint outlines the basis for the alleged violation, abatement period, and proposed penalty. Upon receipt of a copy of the complaint, the employer, within 15 days, must file an answer with OSAHRC. The answer must deny all or portions of the complaint, and admission of guilt is assumed for portions not denied.

SETTLEMENT AGREEMENT

Prior to the case coming before a Commission judge, a formal settlement agreement may be sought. The sequential procedure for establishing a settlement agreement involves the following:

1. An informal conference will be held between the cited firm and OSHA representatives to discuss the compliance parameters desired and felt practical by both parties.

2. A compliance program must be developed by the industry to satisfy the OSHA requirements. Guidelines for compliance plans were discussed previously in this chapter.

3. The compliance program will be submitted to the OSHA Regional or Area Directors.

4. The compliance program will be rejected with guidelines for resubmission or with the decision that the citation provisions will not be modified (this still leaves open legal procedures), or,

5. The compliance program will be accepted, and a settlement agreement will be submitted by the Department of Labor to the cited industry. The settlement agreement will include the following:

a. An outline of the facts of the citation.

b. Reference to the compliance program.

An admission of guilt is implied in the acceptance of a settelment agreement by the respondent. OSHA settlements have no stipulation for dismissal of a case without admission of guilt as do civil or criminal law proceedings.

6. The industry must sign the settlement agreement. This also withdraws its notice of contest.

7. The agreement is signed by the attorneys for the U.S. Department of Labor.

8. The settlement agreement is filed with an Administrative Law judge.

9. The judge sends notice of the settlement to all parties involved, including labor unions.

10. If there is not notice of contest from the notified parties within 15 days, the judge signs the agreement. If a settlement agreement is not attempted or reached, the case will proceed to trial or a hearing before an OSAHRC judge. As complaintant, the burden of proof rests with the Secretary of Labor.

PROGRESS REPORTS

Stipulations of an OSHA noise citation generally require the filing of periodic progress reports with the Area Director. Originally, reports were required every 30 days; however, longer time periods are now allowed. The progress report does not need to be formal or lengthy, but it must provide an accurate summary of progress toward abatement of the cited noise problem.

ABATEMENT PERIOD EXTENSION

The solution of noise control problems is generally not an easy process and is not always achievable by the initial abatement date. When additional time is required, a Petition for Modification of Abatement Date (PMA) may be filed with the Area Director. A copy of this petition must be posted and served to an authorized representative of the employees. The petition should include:

1. All steps taken, and the dates of such action, in an effort to achieve compliance on the requested items during the abatement period.

2. The specific additional abatement time necessary in order to achieve compliance.

3. The reason the additional time is necessary, including the unavailability of professional or technical personnel or of materials and equipment, or because necessary construction or alteration of facilities cannot be completed by the original abatement date.

4. All interim steps being taken to safeguard employees against the cited hazard during the abatement period.

HONESTY

The Act provides that whoever knowingly gives false information is subject to a fine up to $10,000, imprisonment up to 6 months, or both.

INSPECTION CONFIDENTIALITY

One OSAHRC ruling allowed OSHA to revisit a can manufacturing facility to discover additional facts regarding feasibility. The Commission refused OSHA's request to bring in outside experts, however, ruling that only federal employees could be used to protect the company's trade secrets.

EXTENSION OF RESPONSIBILITY

When a new plant is purchased, the new owner is not responsible for a past OSHA citation. He does, however, have a moral obligation to abate known hazards, and his negligence may be considered willful if this obligation is ignored.

LEGAL PRECEDENCE

Decisions handed down by the Occupational Safety and Health Review Commission regarding contested OSHA citations serve to define and clarify the legal responsibilities with which an employer is required to comply. A summary of important OSAHRC decisions is presented in Table 2.2. The complete decision and judicial comment is available as a matter of public record. All OSAHRC decisions relating to noise are summarized in detail in the *Noise Regulation Reporter*, published by The Bureau of National Affairs, Inc., Washington, D.C. 20037.

TABLE 2.2

SUMMARY OF RECENT OSAHRC DECISIONS

Docket	Company	Operation	Decision
1101	Sequoyah Spinningmills	spinners twisters winders	Citation vacated since good faith shown and exposure not severe.
2038	B.F. Goodrich Company	braiders	Abatement required despite estimated cost exceeding \$1 million; Appeals Court remanded and decision not final.
2407	Plaza Beverage Co., Inc.	bottling	Employer not required to use plastic bottles, which was considered only feasible control.
2453	Advance Publications, Inc.	newspaper pressroom	Vacated due to absence of feasible controls.

Docket	Company	Operation	Decision
3707	Anaconda Company	punch press	Citation upheld. No feasible engineering controls established, but employer failed to utilize administrative controls.
3783	Anchor Hocking	glass forming	Failure to prove feasible controls to reduce levels to Table G-16 required vacation of citation.
3827	Reynolds Metals Co.	can mfg.	OSHA did not perform sufficient measurements or identify sources.
4474	International Paper Co.	logging	Feasible controls not available and exposure time not established

Docket	Company	Operation	Decision
4927	U.S. Ring Binder	binders	Noise abatement program justified extension.
5978	Denny Concrete Products Co.	concrete block machine	Feasible controls not proven.
6027	C F & I Steel Corporation	wide flange operation	Vacated; failure to show feasible controls.
6355	Conoco, Inc.	steel bag powder container mfg.	Extension granted despite union opposition due to lack of technology, delays, and progress shown.
6832	Union Timber Corporation	sawmill	Employer requested variance and received instead citation. Vacated because exposure not established, and witness lacked knowledge of sawmills.

Docket	Company	Operation	Decision
6891	Hoerner-Waldorf	sawmill studmill planer resaw finger joint	Vacated; failure to demonstrate significant noise reduction is feasible.
8116	Kaiser Engineers, Inc.	high pressure blow hose	Excessive exposure shown, but hearing conservation provided.
9435	Bonners Ferry Lumber	saws	Exposure calculations upheld.
9616	Seaboard Coast Line RR	locomotive repair	Use of dosimeter as primary measurement device not valid; citation vacated.
10270	Reynolds Metals Company	can forming	Use of dosimeters did not prove overexposure.

Docket	Company	Operation	Decision
10869	Budd Company		It is not the Commission's jurisdiction to review the wisdom of the standard.
10920 11012	Custom Fabrications, Inc.	welding	OSHA survey was inadequate.
10996	Keystone		Extension request upheld.
11043	Continental Screw Company	headers roll threaders	Company to study feasibility of enclosures and install if noise reduction greater than 2 dBA.
11524	Tom Brown, Inc.	drill rig	OSHA failed to prove overexposure with sound meter and dosimeter.
11869	Weyerhaeuser	single facer	OSHA provision of engineering control not valid.

SUMMARY

Summarizing the options related to an OSHA citation, the following sequential procedure may be followed:

1. Inspection.
2. Citation.
3. Informal conference.
4. The employer may admit guilt and proceed with an abatement plan.
5. The employer may file a notice of contest within 15 days.
6. The Secretary of Labor must file a complaint with OSAHRC.
7. The employer must file an answer to the complaint.
8. An informal conference may be held.
9. A settlement agreement may be sought.
10. The case may proceed to a hearing.

Figure 2.1 demonstrates alternative procedures for responding to an OSHA citation.

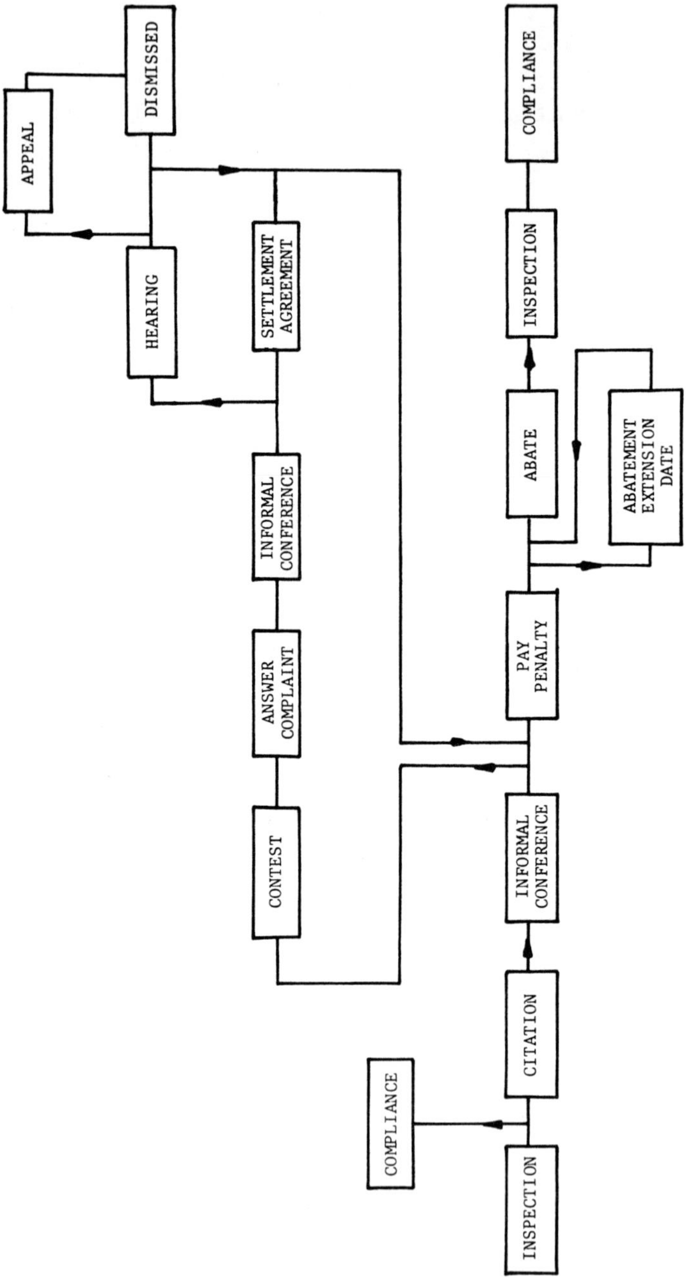

FIGURE 2.1. **Alternatives for response to an OSHA citation.**

Chapter 3
IS IT FEASIBLE?

The Occupational Safety and Health Act requires that engineering or administrative noise control measures be implemented where *feasible*, if noise levels exceed 90 dBA for 8 hours, or a noise dose of unity. There does not, however, exist a special description or definition of "feasible" beyond the Webster dictionary definition, "capable of being done." OSHA enforcement personnel and industry often differ in their opinion of what noise control measures are "capable of being done," and the final determination rests on the legal opinion of an OSHARC judge in many cases.

FEASIBILITY CLASSIFICATION

The difficulty in defining the feasibility status of many problems lies in the misconception that all problems fall into two categories: *feasible* and *non-feasible*. In truth, the noise abatement feasibility of many projects cannot be determined in advance. Unforeseen problems or engineering variables may be encountered in the implementation of any design. Feasibility, therefore, cannot be definitely established until a proposed design is implemented and assessed. Only designs implemented and found to be satisfactory within the industrial environment can be considered to have established their feasibility.

The difficulty of establishing feasibility is well-illustrated by the following case. In an OSAHRC trial, an acoustical expert witness for the U.S. Department of Labor testified that a particular noise control design would provide a feasible means for reducing sound levels for a certain operation. The industry retained the expert and implemented his specific recommendations. The industry was reinspected by OSHA, and a second citation was issued for the same operation! This illustration is made not to imply that the consultant had erred, but to show the difficulty in establishing feasibility prior to actual design and implementation.

Thus, three feasibility categories may be defined:

A. A feasible solution exists for a noise exposure problem for which there exits a noise reduction design approach that:

   (1) does not conflict with production or economic requirements, and,

   (2) has been proven successful in an identical or similar industrial application, or,

   (3) can be determined to be feasible in advance of actual application to the industrial situation based on experimental evidence.

B. A feasible solution cannot be established in advance if the potential solution has not been proven by similar application or extensive experimentation to be successful. Thus, even though a theoretical analysis indicates a noise reduction approach is possible, unless it has been proven by application, it cannot be considered feasible in advance of its implementation.

C. A solution for a noise problem is not feasible if the potential design approaches have theoretical or practical limitations which would preclude effective and successful implementation.

STATE-OF-THE-ART

The OSHA standards do not require that an industry undertake research and development to extend the existing state-of-the-art in search for noise abatement solutions. This limits engineering solutions which are feasible, or required to be implemented, to those defined by the state-of-the-art, or those solutions which have been successfully implemented or otherwise proven to be successful.

This definition should not be considered restricting to OSHA enforcement, since the state-of-the-art in the field of noise control is quite advanced at the present time. Millions of dollars are spent annually on noise control research, and an even higher dollar value is spent on actual implementation of noise reduction measures. With this emphasis on noise control, most noise reduction designs of merit have, indeed, been tried, and are in use somewhere if feasible.

BURDEN OF PROOF

The legal burden of proof in determining a violation of the OSHA standards, and, therefore, of establishing noise abatement feasibility, rests with the U.S. Department of Labor (Title 29, Part 2200, Rules of Procedure, Paragraph 2200.73). In some OSHA offices, it is the policy that a compliance officer should have a potential noise control solution in mind prior to issuing a citation for an operation.

DETERMINATION OF FEASIBILITY

To establish that solution of a noise control problem is feasible, one must consider three areas:

- Acoustical Feasibility
- Production Feasibility
- Economic Feasibility

To establish acoustical feasibility, it must be shown that designs exist which would provide adequate noise reduction.

Each proposed noise control design must be reviewed to insure suitability to the application for which it is intended, and to establish production feasibility. Non-acoustical considerations related to any design include:

1. Employee safety and hygiene.

2. Fire code compliance.

3. FDA compliance for food process plants.

4. Operational integrity:
   a. accessibility to equipment
   b. maintenance servability assurance
   c. product quality assurance

5. Machine system compatibility:
   a. mechanical (power, speed, etc.)
   b. service life
   c. ventilation and cooling

6. Availability of noise control materials.

Although consideration of economic feasibility is not included in the OSHA standard and has not been formally defined by the U.S. Department of Labor, there is evi-

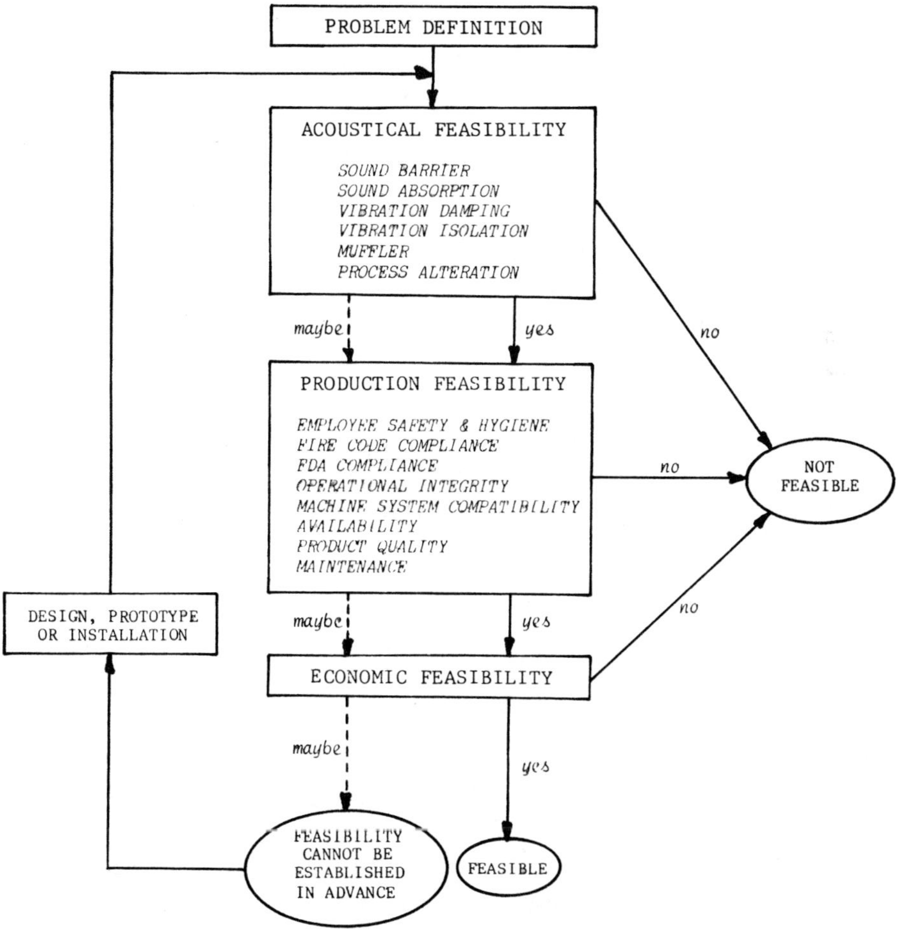

FIGURE 3.1. Decision matrix for determining noise abatement feasibility.

dence that costs are considered in citing alleged violations of the noise code.

Acoustical, production and economic feasibility will be discussed in detail in the following sections. Assessment of each of these feasibility categories must be made in order to determine the *total* feasibility of a proposed noise abatement solution. Figure 3.1 illustrates the matrix of decisions to be made in determining feasibility. In cases where doubt arises as to acoustical or production feasibility, a design prototype may be required.

ACOUSTICAL FEASIBILITY

Fundamental engineering concepts must be reviewed to determine approaches which represent theoretically feasible solutions. Six concepts must be considered. These six concepts include all possible solutions to any noise problem:

a. Sound barriers
b. Sound absorption
c. Vibration damping
d. Vibration isolation
e. Mufflers
f. Process alteration

Design concepts related to theoretically feasible concepts based on the above must be considered. The author, as an acoustical consultant, has utilized the following design investigation procedure to establish a basis for acoustical non-feasibility for several industrial operations:

1. A literature search is performed of all available publications in the noise control field and in the general field of the alleged violation.

2. The problem is discussed with colleagues within the professional community to identify where potential solutions to the problem may have been attempted.

3. Recognized authorities in the academic community are solicited for ideas.

4. The literature of all manufacturers of acoustical materials and systems is reviewed for solution approaches, and many are contacted personally.

5. The manufacturer of the noise-producing equipment is contacted, as are several manufacturers of similar equipment.

6. Trade associations are contacted, and the industry-wide state-of-the-art is sought.

7. Solution approaches are solicited from the OSHA personnel involved in the citation.

PRODUCTION FEASIBILITY

Potential noise control engineering solutions which result in adverse influence on production processes would generally result in higher production costs. Introducing production inefficiencies may also increase the probability of safety problems and injuries.

Safety and Noise Control

Noise control designs which increase the risk of serious physical injury to employees would not be considered feasible. While hearing loss is a serious physical handicap, it is of less magnitude of severity than loss of life or limb. The use of personal protective equipment, although not a failsafe measure, is available as protection against hearing loss. Personal protection may not be available against other possible safety or health hazards that noise control procedures may cause.

There exists no legal system of preemption for conflicting safety and health hazards, and each problem must be judged based on its specific circumstances. Relative priority of engineering control of hazards must be established on the following basis:

a. Severity of hazard.

b. Probability of hazard occurrence.

c. Applicability and availability of personal protective measures.

Noise control engineering measures may adversely affect employee safety in the following manners:

1. Interference with dexterity may be introduced. An individual has an inherent reaction to avoid a potential safety hazard. A noise control structure may be physically located so as to impede the split-second reaction of an individual to avoid the hazard. Any physical restraint of a person's manual dexterity may increase response time in reacting to danger. An example of a noise control system which may constrict operator dexterity is the "glove box", as shown in Figure 3.2. The glove box shields the operator from a noise source, but it also can potentially trap or injure the operator's hand, since it impedes his natural reaction of jerking his hand away from danger.

2. A noise control system may be physically located so that a worker's body may be pinched or trapped between the noise control system and workpiece or other machine.

3. The noise control system may restrict visibility,

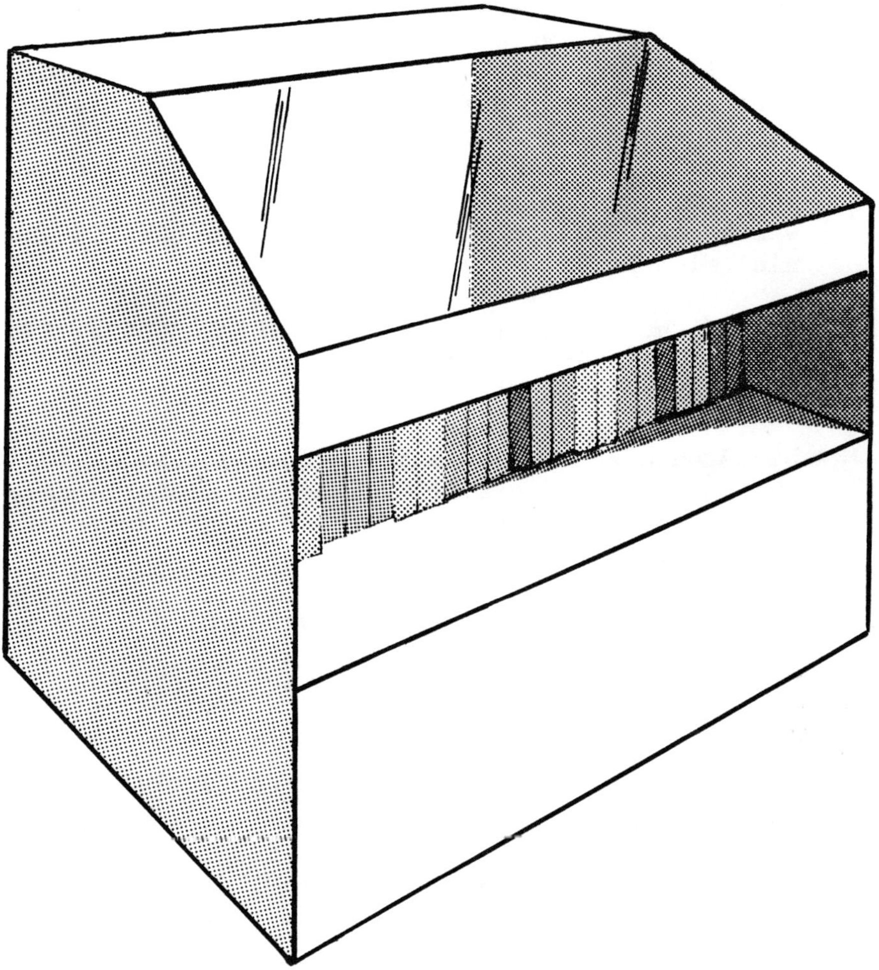

FIGURE 3.2. "Glove box" type enclosure for noise control.

resulting in a potential safety hazard.

4. Noise reduction measures may remove sounds which would serve to warn an employee of a safety hazard associated with his job. It is generally not true that ear plugs or ear muffs would obstruct such warning signals, since a warning signal would be no more attenuated than other noises in the area. However, the following circumstances may occur which would reduce the possibility of hearing warning signals:

   a. If the operator is placed within a booth, air conditioning or ventilation noises may mask operational noise.

   b. A machine enclosure may reduce equipment noise to a level where it will be inaudible in the presence of other background noises exterior to the enclosure.

   c. Machine design and modification of a noise source may reduce or eliminate sounds which employees presently use to assess the machine's operation.

5. Noise control systems may create potential fire hazards.

6. Noise control systems may restrict an employee's mobility which would be required to allow him to move rapidly from a hazard. An example of this is cobbles in steel mills.

7. The machine's strength may be weakened, such as in the case of vibration isolation.

8. The noise control measure may add weight to a tool or machine element, introducing a safety hazard.

9. A toxicity problem may be introduced, such as in the case of some noise control materials which emit toxic vapors when burned.

10. An element of design uncertainty will be present in any production measure which involves designs or devices which have been untried. An unexpected safety problem may be incurred with any such new system.

In general, potential safety hazards which relate to machine failure may be solved through engineering design (i.e., increased structural strength, ventilation, cooling fans, guards, etc.). On the other hand, potential safety problems which relate to employee mobility or dexterity would generally become inherent constraints of an operation, and may not be solvable in any other manner.

Heat Stress and Noise Control

The potential effects due to exposure of employees to hot environments is becoming an area of increasing concern by both industry and the government. It is often found that noise generating machinery also generates heat, resulting in a double hygiene exposure problem. Noise control solutions may either relieve or intensify the heat exposure problem. Where a potential noise abatement solution would increase the effects of heat stress (including heat stroke and even death), noise control may not be feasible.

In many environments, such as steel mills, the use of high velocity air from fans provides the only feasible method to cool workers. These fans, of course, generate noise, but are absolutely necessary to protect employees' health and comfort. The fan noise can be minimized by design but cannot feasibly be eliminated.

Acoustical barriers, walls, enclosures, and partial enclosures may also alter the employee thermal environment, and these designs should be reviewed for their heat stress impact.

Availability of Noise Control Materials

Feasible conceptual approaches to noise problems may exist, but there may not exist any method of implementing these concepts into a production design. A case in point is the impact riveting assembly of large, special design rail cars. The process substitution of squeeze riveting for the impact riveting provides an obvious approach to noise control. However, there are no commercially available or special squeeze rivet hammers of sufficient capacity for this operation. Practical implementation is therefore not feasible.

Despite the abundance of space age materials currently available, acoustical materials suitable for certain specialized acoustical requirements have not been developed. For example, there are no materials which comply with the FDA requirements for intimate food contact, have a Class I fire rating, and posess good sound absorptive properties. Where material or system availability precludes implementation of a noise control concept, alternate designs must be sought or non-feasibility must be recognized.

Product Quality

Some production operations are inherently noisy and involve intimate operator control to insure product quality. For example, the shaping of granite monuments is performed by manual sandblasting. While it is possible to perform this operation automatically, the degradation in monument quality is evident, even to the layman, when the skilled touch of the production artist is removed. Automation, in this case, cannot be considered as an acceptable method of noise control. Numerous similar examples could be cited where isolation of a skilled operator from a production operation would directly and inevitably result in product degradation.

ECONOMIC FEASIBILITY

In the early days of the enforcement of the OSHA noise regulation, the question of economic hardships caused by the OSHA noise standard was given little consideration. More recent sensitivity by governmental officials to industrial reasoning and current economic situations, however, has given rise to consideration of economic feasibility. Incidences such as the layoff of 5000 employees during the peak of the 1974 recession by an Indiana steel mill due to pollution control demands pointed to the need for governmental responsibility for the economic impact of its actions.

If economic feasibility is a valid consideration, and evidence indicates that it is, a legal definition or OSHA guideline is needed. It has been suggested by one industrial trade association that economic feasibiltiy should be clarified within the OSHA standard in well-defined terms, such as:

> "Feasibility" is economically possible if it does not cost more than 10 percent of the cost of direct labor involved on an annual basis, or 10 percent of the cost of the equipment (replacement value), whichever is greater. Reduction of the workday is not considered a feasible control although job rotations shall be so considered.

Charles R. McClure, chief of OSHA's Occupational Health Programming Division, has indicated that both technical and economic feasibility are considered in a noise control project:

> From a technical standpoint, if someone else

"Well, Malcolm, you certainly have proven one thing—that it's feasible to spend $100,000 without lowering the noise level!"

has done it successfully, then at least we can say it is possible. I don't believe, however, that in most cases the problem is lack of technology. There are a few tough situations - such as those found in textile mills and jet-aircraft engine maintenance - but, in the vast majority of the cases, the problem always seems to reduce itself to economics. And that can be a tough nut to crack. What is economically feasible for one company can be entirely infeasible for another. So it really gets down to a case-by-case situation. OSHA has to be convinced of the difficulty of the situation in a given plant before it sanctions the wearing of hearing protection equipment.

NOISE REDUCTION SIGNIFICANCE

It has been the position of the U.S. Department of Labor that engineering noise control measures must be implemented if feasible, whether or not the controls reduce noise exposure to within the OSHA guidelines. It has been the opinion of some industries, supported by OSAHRC decisions, that costly engineering noise control measures are not meaningful or required if noise levels cannot be reduced to a point where the use of hearing protective devices may be eliminated altogether. For in-house assessment of the effectiveness of proposed noise control measures, reduction criteria such as the following might be used:

1. Slight reduction (0-3 dBA)

2. Significant reduction (greater than 5 dBA)

3. Reduction to OSHA guidelines

PRESIDENT FORD'S EXECUTIVE ORDER

The effects of the 1974 recession lead to a review of the economic impact of environmental legislation. Executive Order 11821 issued requirements for consideration of the economic impact of major legislation:

> Hereafter, I will require that all major legislative proposals, regulations, and rules emanating from the Executive Branch include an inflation impact statement that certifies we have carefully weighed their effect on inflation.

AN ECONOMIC ALTERNATIVE

Engineering concepts which are not considered economically feasible by the employer may still be considered technically feasible, and implementation would be required by OSHA. Precedence has been established that noise control expenditures are required as long as the financial stability of a company is maintained. Suppose a feasible technical concept exists for noise abatement which is not economically feasible. Research should be continued to seek a solution which is feasible both technically and economically. In the meantime, of course, interim compliance measures of hearing protection and audiometric testing must be employed, and reasonable engineering progress must be achieved and maintained.

Very little sympathy would be expected by OSHA toward noise violators with costly compliance solutions if good faith has not been demonstrated by protecting workers' hearing by means of a hearing conservation program. Such violations may proceed to legal trial where implementation of an expensive solution may be mandated. Where good faith is shown, it may be expected that considerable time would be allowed to achieve economically reasonable solutions to problems of great magnitude.

ENGINEERING FEASIBILITY STUDY

The engineering feasibility study is probably the most important aspect of an entire noise abatement program, as the results of this study are necessary to establish noise control objectives by management. It is most important that a feasibility study be undertaken prior to engineering design work, since optimum results can only be achieved by implementation of well-established priorities and with consideration of all design alternatives.

If the results of the feasibility study indicate that engineering and administrative controls are not feasible, the OSHA regulation will allow the implementation of a hearing conservation program utilizing personal hearing protectors as a permanent measure of compliance. However, if a feasible solution does exist, the industry is obligated to implement noise abatement measures in lieu of the use of hearing protective devices.

The elements of an engineering feasibility study include:

1. Identification of machinery or operations which contribute to noise exposure in excess of the OSHA guidelines.

2. An outline of potential engineering solutions for each operation.

3. Investigation of acoustical and production feasibility for each operation.

4. An estimate of the noise reduction which may be ex-

pected for each design.

5. Estimate of the cost of each design.

6. Documentation of the basis for non-feasibility where applicable.

Often the services of an acoustical consultant are engaged to provide a wide spectrum of potential solutions and to document the basis for non-feasibility from an independent viewpoint.

Based on the information presented in the feasibility study, management and engineering personnel can establish the following:

1. Priorities for noise abatement based on:
    a. Legal requirements
    b. Benefit-cost ratio
    c. Effectiveness-cost ratio
    d. Severity of hazard
    e. Feasibility

2. Definition of initial basic design approaches from the various alternatives being considered.

3. Engineering design budgets and estimated total project budgets.

4. Documentation of engineering effort as a measure of interim compliance to avoid a possible citation.

5. The basis for an OSHA compliance plan if the company has been cited.

6. Assignment of management and engineering responsibilities for administration, design, supervision, review, implementation, and assessment.

REFERENCES

1. Meinhold, T. F., "About Those Hearing Protectors," *Plant Engineering*, Vol. 29, No. 18, September 4, 1975, pp. 68-71.

Chapter 4
SURVEY FOR RESULTS

The OSHA regulation only briefly defines measurement requirements by stating[1]:

> Protection against the effects of noise exposure shall be provided when the sound levels exceed those shown in Table [G-16] of this section when measured on the A scale of a standard sound level meter at slow response.

The U.S. Department of Labor has further defined noise survey requirements as follows[2]:

> A noise survey of each area in the plant in which sound levels exceed 90 dBA shall be made at least once each year to insure that sound levels have not increased above those originally existing. The survey may also establish that noise levels in some areas have been reduced to levels below 90 dBA and thereby justify discontinuing application or requirements for administrative controls, ear protection, and audiometric tests of individuals in such areas.
>
> A noise survey of an area is recommended whenever a change is made in either equipment or type of operations so that significant changes in noise level will be acted upon immediately.
>
> Tests of noise levels will be made with a sound level meter on the A-scale, slow res-

ponse. The use of octave band analyzers or impact meters for control or other purposes shall be in addition to, not in place of, tests made with the sound level meter.

The sound level meter used will be one meeting specifications in ANSI Standard S1.4-1961, "General Purpose Sound Level Meters."

Records will be made of such surveys showing: the instrument used; date, time and location of such tests; machinery or equipment generating the noise; and name of person making the test.

Test records shall be kept readily available for inspection for 1 year or until a subsequent survey is made, if done more frequently.

The noise survey will be made by an insurance carrier, a consultant, a representative of a state health or labor department, or by a qualified individual designated by the company.

PERFORMING THE CORRECT MEASUREMENTS

Heading the list of wasted noise abatement dollars are misguided noise measurements efforts. The errors which are incurred in the management of a noise measurement program may be classified into five categories:

A. *Unnecessary Data*. Excessive measurements may be performed beyond those required by the objectives of a noise survey. How many times have octave band sound pressure levels been measured to assess OSHA compliance, while the regulation criteria require only A-weighted sound level measurements? Exces-

"Yes Wally, but are you sure we need all this sound instrumentation?"

sive measurements waste money!

B. *Inadequate Data.* Acoustical designs may be approached without adequate measurement data. Many times designs are based only on A-weighted sound level measurements. This procedure can result in an optimum design only if this single-number data is supplemented by a stroke of luck!

C. *Unnecessary Instrumentation.* Measurement instrumentation is purchased without establishing measurement objectives. How often is instrumentation purchased and never actually used in solving the noise problems for which it was intended?

D. *Erroneous Assessment.* Measurement instrumentation exceeds the capabilities or training of the investigator. The OSHA measurement requirements have been designed such that noise exposure assessment may be performed by individuals without a strong technical background. However, when more specialized measurements are performed by individuals without specialized training, costly errors in assessment and design recommendations may result.

E. *Superfluous Assessment.* In-depth noise exposure assessment measurements are performed by the consultant or in-house engineer beyond those required to establish legal compliance or define hearing damage risk. In some cases, effort and expenditure in a noise survey is excessive to the point where the same level of effort, if directed toward engineering design, could have solved the problem.

DEFINITION OF PURPOSE

To avoid errors in the instrumentation used and the type of measurements performed, it is necessary to determine in advance the purpose of the study. In order to categorize measurement requirements, it is necessary to define the acoustical project objectives. The three general measurement objectives are: *legal compliance* requirements, *hearing conservation* requirements, and *engineering design* requirements.

LEGAL COMPLIANCE REQUIREMENTS

The OSHA regulations clearly define sound environments which require corrective procedures in terms of A-weighted sound levels and employee exposure times. Recognizing sound levels which exceed these criteria is relatively simple, and generally requires only brief measurements. For the sole purpose of ascertaining that legal noise control obligations exist, it is generally unimportant to perform measurements of extreme accuracy. For example, it may not be relevant to distinguish between a noise dose of 180% and one of 250%, since both clearly exceed the OSHA guidelines. Acoustical measurements beyond simple dBA sound level readings are generally not necessary solely for the purpose of establishing legal compliance. Only for the case where legal compliance is marginal (i.e., sound levels very near 90 dBA or noise doses near 100%) would detailed sound level or exposure time data be appropriate. Also, where actual exposure time is not evident, such as in the case of intermittent noise sources, detailed dosimeter or time study measurements should be performed.

HEARING CONSERVATION REQUIREMENTS

Hearing damage risk must be assessed with a criteria other than the OSHA regulations, since permanent hearing loss may result from exposure to sound levels less than 90 dBA. Sound level thresholds of 55 dBA, 70 dBA, and 80 dBA have been identified by various investigagators as presenting hearing damage risk. These criteria are discussed in Chapter 5. Sound level measurements should be performed in all cases where sound level exposures are above the recognized threshold. The sound level data should have an accuracy and reliability sufficient to provide effective documentation in he defense of a worker's compensation hearing loss claim. The assessment should also define the hearing conservation action which is appropriate for each employee's noise exposure. Hearing conservation action levels may be classified into six exposure categories:

I. *Below 55, 70 or 80 dBA* (whichever is recognized as hearing loss threshold)*:* Baseline audiometric test only.

II. *55, 70 or 80 dBA to 85 dBA:* Baseline audiometric test given periodically (usually less frequent than annually). Hearing protection furnished upon employee request.

III. *85 dBA to 90 dBA:* Baseline audiometric test given annually. Hearing protection furnished upon employee request or mandatory for employees whose audiograms indicate hearing degradation.

IV. *90 dBA to 105 dBA:* Baseline audiometric test given annually and implementation of engineering and/or administrative controls, with mandatory hearing protection until sound levels are reduced.

V. *105 dBA to 115 dBA:* Same as IV, with special assessment of hearing protection requirements.

VI. *Above 115 dBA:* Same as V, with special audiological or medical study or potential physiological or psychological hazards due to noise exposure.

ENGINEERING DESIGN REQUIREMENTS

The development of engineering recommendations and designs requires advanced measurements. These include, at a minimum, octave band sound pressure level measurements, and often vibration analysis or narrow band frequency analysis. Specific requirements will vary with the type of noise problem encountered. For a technical description of acoustical instrumentation and measurement procedures, the reader is referred to References 5 through 8.

SURVEY PROCEDURE

The sound survey procedure developed by General Motors Corporation is one which clearly defines OSHA compliance without overassessing the problem. Their procedure is defined as follows:

> The purpose of the Sound Survey Procedure is to provide a uniform and efficient method for determining employee noise conditions in each General Motors plant. It also establishes instrumentation requirements and data reporting practices. The survey procedure is a three-step process as shown in the Sound Survey Logic Diagram [Figure 4.1].
>
> *Phase 1*. Using a sound survey meter set for A-weighting, slow response, the regularly occurring maximum noise level and the regularly occurring minimum level are recorded at the center of each bay. (Bay size is limited to 1000 sq. ft. or smaller.) If the maximum sound level does not exceed 84 dBA, it is assumed that all employees in this area are working in a satisfactory noise atmosphere. If the minimum sound level measured at center of the bay

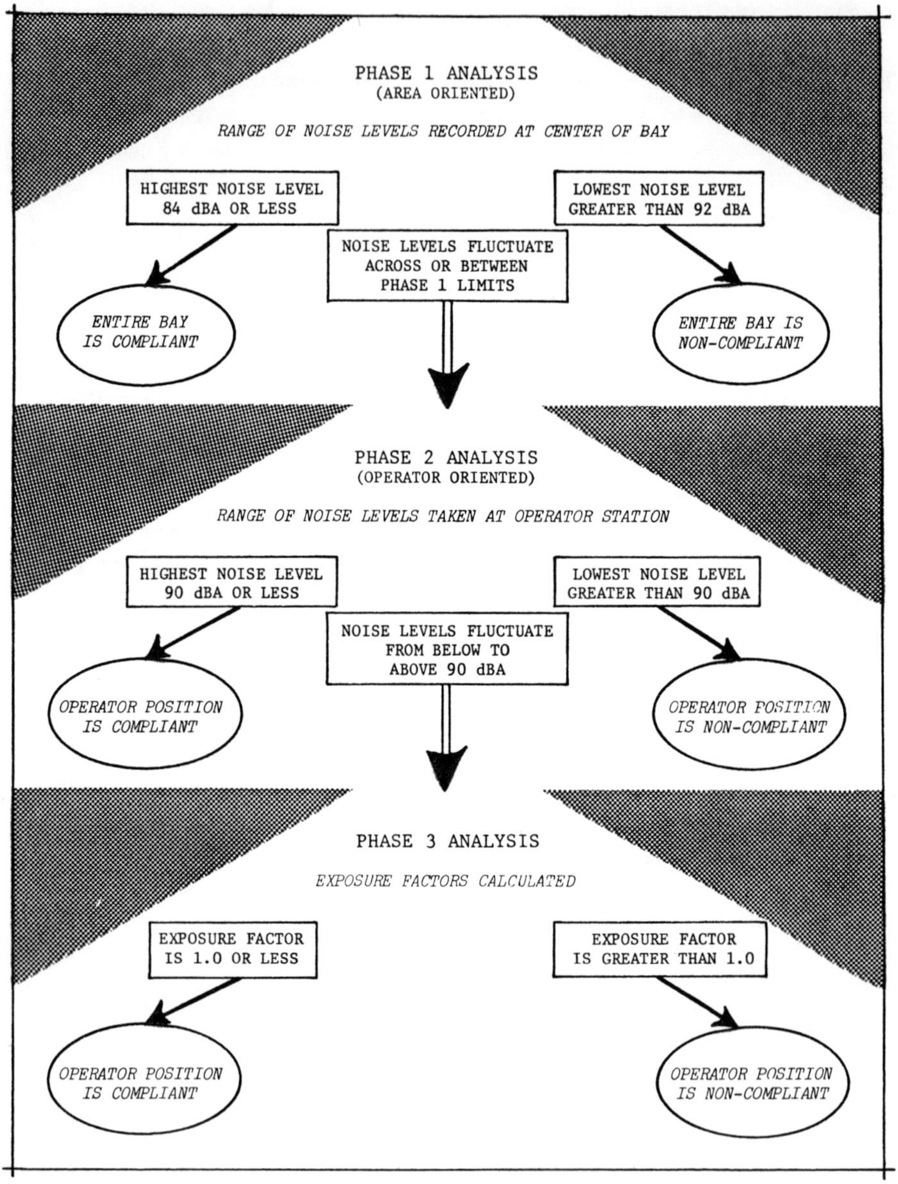

FIGURE 4.1. Sound survey logic diagram.[4]

is 92 dBA or higher, then it is assumed that all the employees in this area are working in an unsatisfactory noise atmosphere. If the range of levels measured at the center of the bay falls between 92 and 84 dBA, then more information is needed before individual employee exposure can be determined. Up to this point, the two types of plant areas have been classified as being satisfactory or unsatisfactory, using a minimum of time and expense. Now it becomes necessary to evaluate the noise exposures for people working in bays where measurements at the center of the bay range from 84 to 92 dBA. These measurements are too near the permissible limit of 90 dBA to estimate; therefore, readings must be made at the employee's position.

Phase 2. Sound level measurements are made at the operator's position. Again, if the level varies on a regular basis, both the maximum and minimum levels are recorded. Work station measurements are much easier to evaluate because if the noise level never goes below 90 dBA, the cumulative exposure must be greater than unity, which indicates an unsatisfactory employee noise situation. Conversely, if the measurement is never above 90 dBA, the cumulative exposure must be less than unity and the employee station can be recorded as satisfactory.

Phase 3. At work stations where the regularly occurring noise varies from above 90 dBA to levels below 90 dBA, Phase 3 analysis must be used. This simply means that noise exposure factors based on noise level and time must be calculated for this situation. At the conclusion of Phase 3 analysis, the surveyor will be able to determine, for each employee, whether the employee is working in a hazardous or nonhazardous noise environment. At this point, the plan of attack can be organized and the job or protecting employees begins.

SUMMARY

The following guidelines should be remembered in a noise survey program:

1. Measurements do not save employee hearing. They provide information for assessment and engineering design, and should be performed when necessary for these purposes. A noise survey is a first step, not an end result in a noise program.

2. For sound level assessment for legal compliance and hearing conservation programs, octave band analysis is generally not necessary.

3. For design of engineering noise control measures, frequency analysis such as octave band sound pressure level measurements or narrow band analysis are necessary to insure the development of effective, optimized solutions.

REFERENCES

1. Williams-Steiger Occupational Safety and Health Act of 1970, PL91-596, December 29, 1970.

2. "Guidelines to the Department of Labor's Occupational Noise Standards," Bulletin 334, Revised, 1971.

3. American National Standard Specification for Sound Level Meters, ANSI S1.4-1971.

4. Van Tifflin, W. L., "GM's Noise Control Program," Proceedings of the Technical Program, NOISEXPO, September 11-13, 1973.

5. Peterson, A. P. G., and Gross, E. E., *Handbook of Noise Measurement*, General Radio, 1972.

6. Thumann, A., and Miller, R. K., *Secrets of Noise Control*, Second Edition, The Fairmont Press, Inc., 1976.

7. Keast, D. N., *Measurements in Mechanical Dynamics*, McGraw-Hill, 1967.

8. Broch, J. T., *Acoustic Noise Measurements*, Bruel & Kjaer, 1971.

Chapter 5

REDUCING RISK OF WORKER'S COMPENSATION PAYMENTS *

It has been well established that exposure to high noise levels results in permanent loss of hearing. The extent of loss is dependent upon the sound intensity, exposure duration, and susceptibility of an individual.

A recent interagency study between the U.S. Air Force and U.S. Environmental Protection Agency resulted in the preparation of three comprehensive documents[1,2,3] on noise exposure. The relationship between continuous noise exposure and hearing loss was identified in this study as a result of information compiled by four investigators[4,5,6,7]. These results are shown in Figure 5.1. For a detailed summary of the results of the interagency study, the reader is referred to Reference 8.

Subsequent to the interagency study, the U.S. Environmental Protection Agency proposed a 24 hour equivalent noise level of 70 dBA be designated the limit requisite to protection against hearing loss with an adequate margin of safety[9].

The U.S. Department of Labor recently published results summarized from various professional studies which identified hearing loss resulting from long term exposure to various sound levels[10]. The percent of popu-

---

* The assistance of Hearex Corporation, Tampa, Fla. in providing information for this chapter is appreciated.

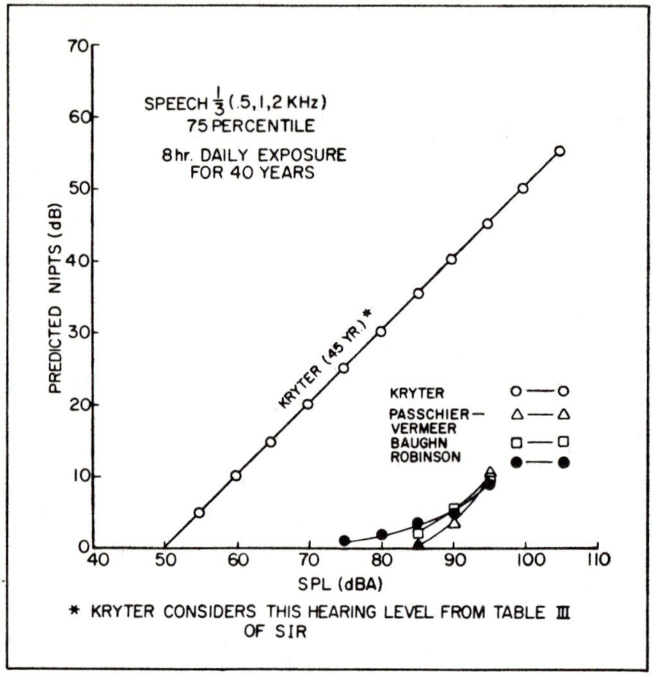

FIGURE 5.1. Noise-induced permanent threshold shift (NIPTS) of the traditional speech frequencies (.5, 1, and 2 kHZ), for 40 year exposure on an 8 hour daily basis, as predicted by four different researchers.[8]

TABLE 5.1

PERCENT OF POPULATION EXHIBITING MORE THAN 25 dB HEARING LEVEL RE ISO ZERO AFTER 40 YEARS OF EXPOSURE TO SPECIFIED NOISE LEVEL[10]

Noise Level	Total %	% Due to Noise	% Due to Other
80	24	0	24
85	32	8	24
90	42	18	24
95	52	28	24
100	64	40	24
105	78	54	24
110	88	64	24
115	94	70	24

Note: ISO ZERO is defined as the hearing threshold for audiometric testing by the International Standards Organization.

TABLE 5.2

INCOME BENEFITS FOR HEARING LOSS

Jurisdiction	One Ear	Both Ears
Alabama	$ 3,604	$ 11,084
Alaska	3,640	14,000
Arizona	11,000	33,000
Arkansas	2,520	9,450
California	2,100	10,500
Colorado	2,266	9,000
Connecticut	5,824	17,472
Delaware	5,625	13,125
District of Columbia	10,948	42,108
Federal Employees' Compensation Act	27,000	103,846
Florida	3,200	12,000
Georgia	3,900	9,750
Guam	2,912	11,200
Hawaii	5,850	22,500
Idaho	2,603	11,157
Illinois	4,845	12,112
Indiana	4,500	12,000
Iowa	4,200	14,700
Kansas	1,680	6,160
Kentucky	6,225	12,948
Longshoremen and Harbor Workers' Act	10,948	42,108
Louisiana	-	-
Maine	4,150	16,600
Maryland	7,275	22,601
Massachusetts	4,500	12,000
Michigan	-	-
Minnesota	5,500	17,000
Mississippi	2,240	8,400
Missouri	2,860	10,920

TABLE 5.2 (Continued)

Jurisdiction	One Ear	Both Ears
Montana	$ 2,400	$ 12,000
Nebraska	4,000	8,000
Nevada	-	-
New Hampshire	5,980	24,610
New Jersey	2,400	8,000
New Mexico	2,600	9,750
New York	4,800	12,000
North Carolina	5,600	12,000
North Dakota	2,000	8,000
Ohio	1,802	9,012
Oklahoma	5,000	10,000
Oregon	4,200	13,440
Pennsylvania	5,640	24,440
Puerto Rico	2,250	9,000
Rhode Island	2,700	9,000
South Carolina	4,851	10,395
South Dakota	-	-
Tennessee	-	-
Texas	4,725	9,450
Utah	4,450	8,900
Vermont	4,212	17,415
Virginia	4,000	8,000
Washington	2,400	14,400
West Virginia	5,520	16,560
Wisconsin	2,915	17,490
Wyoming	-	-

Note: Table based on U.S. Chamber of Commerce Analysis of Workmen's Compensation Laws (January 1974).

lation expected to exhibit a significant hearing loss due to various noise levels with 40 years of exposure is given in Table 5.1. It will be noted in the table that 24% of the population will incur a hearing loss due to causes other than noise exposure. Thus, the total population incurring a hearing loss will include both those exposed to noise and the non-noise-exposed group.

The loss of hearing was recognized as compensable in 1948, when a claim was awarded in the case of Slawinski versus J. H. Williams and Company[11]. Presently, all states either have provisions for compensation for occupational noise induced hearing loss, or have such proposed or pending legislation. Table 5.2 presents a status report of state hearing loss compensation policies. Any such list will become outdated rapidly, as new and revised legislation appears almost weekly with the present emphasis on hearing loss.

Hearing loss claims are approaching the crisis level for many industries, and in many states. Recently a $30,000 compensation award was ordered for a single individual hearing loss. Some plants have reported an annual payment in excess of $3,000,000.

One recent article, which recognized that not all workers will suffer severe losses or will apply for compensation, estimated the average compensation payment to be $2000. Governmental and insurance company estimates indicate approximately 6,000,000 workers are affected by noise environments which may result in hearing loss. With these estimates, the cost to industry would be $12,000,000,000!

Not only are hearing loss claims increasing in amount, but the frequency of claims is increasing. For example, the state of Oregon reports the following statistics:

"It's great! Now that I've won my $8,000 workmen's compensation suit for hearing loss, I can get a snowmobile and replace my old cycle and electric guitar!"

Year	*Hearing Loss Claims*
1971	2
1972	25
1973	700
1974	1500

Payment of worker's compensation payments for hearing loss is not an obscure potential future problem for industry. It is a *real* and *present* problem. A survey of 15 industrial insurance companies shows that hearing loss is the largest single compensable health problem today, ranking above black lung disease and exposure to vinyl chloride.

To date, filing of hearing loss claims has been more prevalent in certain states and regions than others. Industries in New York, New Jersey, Oregon and California have been the hardest hit by claims. One plant has reported that claims totaling $12 million dollars have been filed by its employees. Although settlements are generally much smaller than the amount sought, the dollar value for compensation is geometrically increasing.

Public awareness of occupational hazards and hearing loss risk has been largely responsible for employee action to seek compensation. Newspaper and magazine articles frequently discuss occupational hazards. One recent article, for example, was entitled, "Your Job May Be the Death of You." Trade unions have developed films and publications on hearing loss and often provide guidance to the employee in filing a claim.

Quite naturally, industry is not pleased about the billions of dollars in potential claims. It is a common view of industrial executives and safety managers that employees who suffer disabilities related to their job are, indeed, entitled to appropriate compensation. It is not the intention of industry to deprive their employees of just compensation. The major concerns of industry regarding worker's compensation are:

1. How can abuses, inconsistencies and errors in the compensation system be reduced?

2. How can hazards which result in compensation payments be prevented?

Industry has two fortunate advantages in combating unjust claims and reducing the hazard risk:

1. Hearing loss can be prevented.

2. Hearing loss, or absence of a loss may easily be documented, and an industry may avoid payment of claims for hearing loss for which it is not responsible by this documentation.

Unfortunately, the available methods of compensation prevention have not been employed by a majority of industrial plants. The potential cost of claims provides more than sufficient financial incentive for industry to implement measures to prevent hearing loss. A most obvious initial approach is the use of hearing protective devices. While a program of protection may significantly reduce hearing loss, audiometric testing is necessary for actual determination of the effectiveness of the hearing protection. Without audiometric baseline and follow-up tests, hearing loss protection provided by personal protective devices is difficult to establish in possible worker's compensation litigation.

THE HEARING CONSERVATION PROGRAM

Since the inception of the Occupational Safety and Health Act of 1970, Section 1910.95 of the regulation has required[12]:

> In all cases where the sound levels exceed the values shown herein, a continuing, effective hearing conservation program shall be administered.

U.S. Department of Labor Bulletin 334 defines "hearing conservation" as follows[13]:

> In a broad sense, "hearing conservation" covers the entire range of actions required by sections 50-204.10 and 1910.95. As used here, however, it refers to audiometry - periodic checks on the hearing ability of individual employees; and noise surveys - periodic checks of the noise level in the areas in which employees are working.

Reference 13 describes the specific U.S. Department of Labor requirements for hearing conservation. The revisions to the OSHA noise standard, presented in Appendix B, provide updated guidelines.

AUDIOMETRIC TESTING

The only method for obtaining reliable documentation to define the hearing acuity of an individual is through testing on reliable and calibrated audiometric test equipment. The revised OSHA noise standard (Appendix B) outlines procedural requirements for performance of reliable audiometric testing.

An audiometric test is the *only* recognized measure which can protect a company against payment of worker's compensation claims for hearing loss. To disclaim responsibility for an employee's hearing loss, a company must produce evidence that the loss was not caused by his occupational exposure. This is easily accomplished by reference to audiometric test records which document

TABLE 5.3

IN-HOUSE AUDIOMETRIC TESTING FIXED COSTS

Equipment	1st Year	Subsequent Years
Test Booth	$1,300	-
Audiometer	500	-
Sound Level Meter	500	-
Otoscope	50	-
Certification	225	$100
Annual Calibration (meter & audiometer)	100	100
Disposables (otoscope tips, alcohol, kleenex, etc.)	50	50
Publications & Conference on Hearing Conservation	150	150
Audiograms (for under 1000)	50	50
Ear Muffs (25 @ $8 for 14 hour pre-test quiet)	200	-
TOTAL ANNUAL COST	$3,125	$450

hearing loss prior to employment or due to non-occupational exposure. If an industry doesn't have documentation of an employee's hearing loss prior to employment, it may be required to pay compensation for the loss.

The purpose of an audiometric test must be recognized; an audiometric test does not directly prevent hearing loss. Only hearing protective devices or other noise exposure reduction measures will *prevent* hearing loss. Audiometric tests only provide *documentation* and *proof* that hearing protection measures are effective in preventing hearing loss. Conversely, hearing protective measures cannot reliably be assumed to be protecting an employee's hearing. There are several reasons why hearing protective devices may not adequately protect hearing, including improper fit, improper application, and lack of use. Audiometric verification is necessary to prove that the devices are effective for each individual using the personal protection.

CONTRACT VS. IN-HOUSE AUDIOMETRIC TESTING

Employee audiometric tests may be performed by in-house personnel or by an independent firm. These testing firms generally utilize a mobile test laboratory and perform audiometric evaluation on a contract basis. The two approaches may be compared with regard to three criteria:

a. Cost
b. Administrative and manpower responsibility
c. Professional responsibility

Table 5.3 summarizes cost requirements for a typical in-house audiometric testing program. Table 5.4 provides a comparison between the basic costs of in-house and contract services. While costs may vary slightly

TABLE 5.4

COMPARATIVE COST OF IN-HOUSE AND CONTRACT AUDIOMETRIC TESTING

	IN-HOUSE		CONTRACT	
No. of Employees to be Tested	First Year*	Following Years*	Annual	
100	$ 3,675	$ 1,000	$ 650	(@ $6.50)
200	4,225	1,550	1,300	"
300	4,775	2,100	1,950	"
400	5,325	2,650	2,600	"
500	5,875	3,200	3,000	(@ $6.00)
600	6,425	3,750	3,600	"
700	6,975	4,300	4,200	"
800	7,525	4,850	4,800	"
900	8,075	5,400	5,400	"
1,000	8,625	5,950	5,500	(@ $5.50)
1,100	9,175	6,500	6,050	"
1,200	9,725	7,050	6,600	"
1,300	10,275	7,600	7,150	"
1,400	10,825	8,150	7,700	"
1,500	11,375	8,700	8,250	"
2,000	14,125	11,450	11,000	"

* Based on fixed cost (Table 5.3) plus the following costs per audiogram: Nurse ($4.); Doctor ($1.); Record Keeping and Storage ($.50).

for various plants, values of the table may be adjusted to provide a specific analysis for a particular plant.

Administrative requirements of an audiometric testing program include the following:

1. The technicians must be trained, certified and supervised.

2. Audiometric tests must be scheduled to provide minimal interruption of production.

3. Testing instrumentation must be acquired, calibrated and stored.

4. Audiometric test results must be evaluated and stored.

Professional requirements of an audiometric testing program include the following:

1. The program should be supervised by a medical doctor or audiologist.

2. Audiograms should be reviewed by a medical doctor or audiologist.

3. The technician administering the tests should be certified and must be appropriately insured.

4. The administering technician and supervisors should not have a vested interest in the test results.

5. Instrumentation and test environments should meet or exceed all applicable standards.

The importance of professionally administered tests cannot be overemphasized, as this will be the primary factor in determining the success of a defense in potential worker's compensation hearing loss cases. It is of little benefit for an industry to invest the time and cost for an audiometric program if the results will not meet the professional requirements necessary to prevent compensation claims.

HEARING PROTECTIVE DEVICES

Sound levels reaching the inner ear may be effectively attenuated by the use of hearing protective devices. Two categories of devices are commonly used for industrial hearing conservation:

1. *Aural.* Insert type "ear plugs" protect by blocking the external ear canal. These devices may be of the pre-formed, molded type, or shapeless, moldable type.

2. *Circumaural.* "Ear muffs" utilize cups placed over the ear to block sound before it reaches the external canal.

Typical attenuation values of various types of hearing protection are presented in Table 5.5. In general, most commercially available devices will provide effective attenuation for sound levels up to 105 dBA; many devices will provide an even greater attenuation.

A recent publication by NIOSH, "List of Personal Hearing Protectors and Attenuation Data," lists attenuation data for 175 ear protectors by 40 manufacturers. It is designated HEW Publication No. (NIOSH) 76-120, and is available from:

NIOSH
Office of Technical Publications
Post Office Building, Room 530
Fifth and Walnut Streets
Cincinnati, Ohio 45202
(513) 684-2723

TABLE 5.5

TYPICAL ATTENUATION VALUES, IN DECIBELS, OF HEARING PROTECTIVE DEVICES

Type	FREQUENCY, HZ						
	125	250	500	1000	2000	4000	8000
Ear Muff	15	22	33	44	43	48	32
Plug, Custom	18	16	17	20	35	44	35
Plug, Prefabricated	26	25	26	28	35	41	40
Plug, Disposable	25	25	25	26	35	40	37
Muff & Plug	37	35	39	45	44	49	40
Waxed Cotton	16	19	22	26	33	44	36
Cotton	-	3	5	9	13	13	16

The selection of specific types of devices generally depends on:

a. Performance
b. Comfort
c. Application
d. Appearance
e. Cost

Many plants find a greater employee acceptance if several alternate types of protection are offered.

The use of personal protective equipment is considered by the Department of Labor to be an interim compliance measure while engineering and administrative controls are being perfected. There will be very few cases in which the use of this equipment will be acceptable as a permanent solution to noise problems.

It is generally accepted that hearing protective devices prevent hearing loss if they are properly worn and the sound fields are not extreme. The reasons that hearing protective devices are not allowed as a permanent compliance measure by OSHA in lieu of engineering and administrative controls are:

1. There is evidence of non-auditory health hazards due to noise against which personal hearing protective devices are ineffective.

2. Hearing protective devices must be properly worn and fitted to be effective.

3. Hearing protective devices may cause ear infections if not kept properly clean.

4. Employees do not always wear hearing protective devices when issued.

Hearing protective devices should be issued by an experienced nurse or technician who instructs the employee on proper wearing and cleaning. The devices should be individually fitted, and alternate types should be offered if the employee experiences a comfort problem.

The wearer's ear should also be inspected by an experienced technician or nurse to detect signs of infection or other disorders. The wearing of ear plugs can irritate an ear infection. If ear plugs are kept clean, however, they will *not* cause an ear infection.

Two types of programs may be implemented to encourage employees to wear their issued hearing protection:

(a) *positive incentive program*
(b) *negative incentive program*

A positive incentive program involves educating plant employees on the merits and benefits of wearing the hearing protection. The incentive program could include competition between departments. For example, a reward may be given to the department which has the best record of properly worn protectors.

A negative incentive program entails issuance of the protection on a mandatory basis. Employees who fail to wear the devices are given penalties, which may include verbal warnings, written warnings, suspensions, and severance of employment.

RESULTS OF A HEARING CONSERVATION SURVEY OF 4000 COMPANIES[12]

In June 1975, NIOSH published the results of a survey of 4000 selected companies representing manufacturing, construction, transportation and mining industries. The purpose of the study was to:

1. Find on-going industrial hearing conservation programs.

2. Assess the extent and nature of the variability in these existing conservation efforts.

3. Learn the range of problems involved in establishing such programs.

4. Determine apparent levels of effectiveness.

Approximately 2000 companies returned the questionaire. Of these, 44% claimed to have noise problems, 27% had noise reduction programs, and 23.5% had programs concerned with hearing conservation.

The majority of industries surveyed indicated that the thrust of their hearing conservation effort was the use of engineering procedures to reduce noise exposure. Noise control at the source posed the greatest problem. In many instances there were not enough funds to implement engineering controls and technical expertise was lacking. Administrative controls had seldom been used with much success. This was primarily due to complications which arose when a worker had to be shifted to a job which had a different pay rate classification, and objections raised by the unions. Personal ear protectors were the most widely used single method for controlling employee exposure to noise. These programs were reported to be most effective when proper use, care, and importance of wearing the ear protectors had been demonstrated to the employees. Improved employee cooperation occurred in cases where a wide variety of personal ear protectors were offered, and a trial period was given to the employee to permit him to select the type he liked best.

The NIOSH survey found that most firms with audiometric testing programs supplemented this effort with hearing protective devices, noise surveys and noise control en-

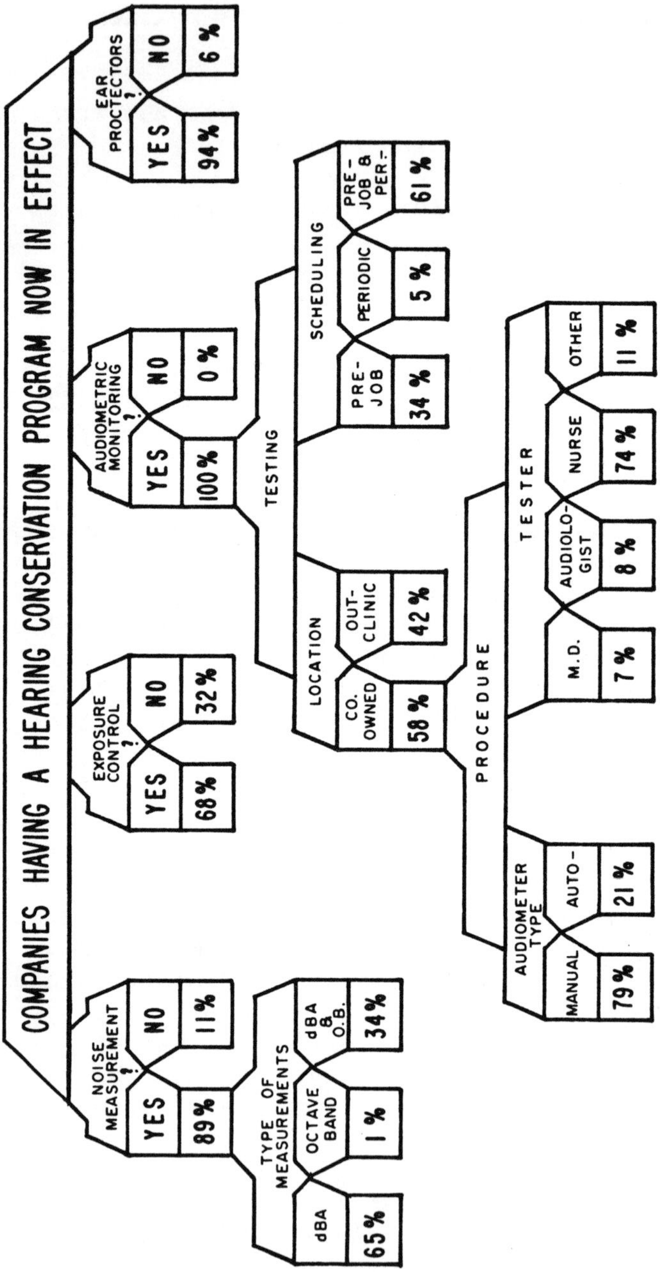

FIGURE 5.2. Summary of hearing conservation practices.[13]

gineering. Most firms performed both pre-employment and periodic tests, and both in-plant and contract (out-clinic) programs were popular. Most in-plant audiometric tests were administered by a nurse with a manual audiometer. A summary of the statistics associated with these programs is presented in Figure 5.2. In surveying the practices of several companies, it was found that the calibration of 80% of the in-plant hearing testing facilities failed to meet the ANSI criteria.

Based on a sample of 430 plants, or 2.7% of the national workforce, it was found that 14.9% of the employees were located in areas where sound levels were 90 dBA or above. This data is presented in Table 5.6. A breakdown of the noise exposure by industry classification is shown in Table 5.7.

Most companies had performed noise surveys, and it was found that a general purpose sound level meter with an A-weighted network was used in 68% of the surveys (see Figure 5.3). Octave band sound pressure level readings, generally with A-weighted readings, were used in other cases. A breakdown of the training of the survey personnel is illustrated in Figure 5.4.

Engineering controls were reportedly tried by almost every company. The engineering procedures ranged from replacing worn machinery parts to changing processes. The main complaints about using engineering controls were the difficulty in finding qualified consultants, the uncertainty of available technology, and the expense. The costs ranged up to $40,000. Companies also expressed the difficulty in accurately predicting the resulting attenuation from proposed engineering controls. In one case a $35,000 investment for quieting one noise source reduced the overall noise level by only 2-3 dBA.

Disciplinary actions for not wearing ear protectors

TABLE 5.6

NUMBER OF EMPLOYEES REPORTED TO BE EXPOSED TO NOISE LEVELS OF 90 dBA OR ABOVE

	Number of Plants Included in Sample	Total Number of Employees in Sample	Number Located in Areas 90 dBA or Above	Estimated % of Workforce Exposed*	Total Work Force	No. Projected to be in Areas 90 dB
Manufacturing	351	513,166	72,188	14.9	16,546,400	2,470,681
Construction	6	755	82	10.9	3,480,000	377,960
Transportation	24	86,640	16,936	19.5	2,193,600	427,752
Mining	49	26,455	3,994	15.0	378,545	56,732
TOTAL	430	627,016	93,200	14.7	22,598,545	3,333,125

* These percentages are calculated by dividing Total Work Force into No. Projected.

TABLE 5.7

NUMBER OF EMPLOYEES REPORTED TO BE EXPOSED TO NOISE LEVELS OF 90 dBA OR ABOVE IN MANUFACTURING

SIC	CODE	No. of Plants in Sample	Total No. of Empl. in Sample	No. Located in Areas 90 dBA and Above	Percentage of Work Force Exposed in Sample	Total Work Force	No. Projected to be Located in Areas 90 dBA and Over
22	Textile Mill Products	23	12,764	5,634	44.1	963,300	424,815
29	Petroleum & Coal Products	17	24,193	6,675	27.6	192,800	53,212
24	Lumber & Wood Products	14	5,654	1,460	25.8	601,000	155,058
20	Wood & Kindred Products	16	22,433	5,769	25.7	1,898,600	487,940
25	Furniture & Fixtures	11	10,374	1,849	17.8	465,400	82,841
34	Fabricated Metal Products	58	41,428	7,079	17.1	1,335,000	228,285
32	Stone, Clay & Glass Products	5	2,502	416	16.6	643,800	106,870
33	Primary Metal Industries	54	74,736	11,081	14.8	1,190,000	176,120
30	Rubber & Plastic Products	5	8,691	1,105	12.7	589,500	74,867
37	Transportation Equipment	46	199,212	23,445	11.8	1,705,500	201,249
36	Electrical Equip. & Supplies	7	8,790	973	11.1	1,778,100	197,369
28	Chemicals & Allied Products	9	3,851	324	8.4	1,014,400	85,210
23	Apparel & Other Textile Products	1	50	5	10.0*	(1,353,100)	
26	Paper & Allied Products	22	15,277	1,400	9.2	687,300	63,240
19	Ordnance & Accessories	12	39,403	3,480	8.8	193,900	17,063
38	Instruments & Related Products	6	3,254	193	5.9	433,800	25,594
35	Machinery Except Electrical	39	25,327	1,043	4.1	1,768,000	72,488
27	Printing & Publishing	6	15,227	257	1.7	1,085,900	18,460
	TOTAL:	351	513,166	72,188	**	16,546,400	2,470,681

* Insufficient data for projection

** Based on No. Projected and Total Work Force, the estimated percentage of total work force exposed is 14.9%

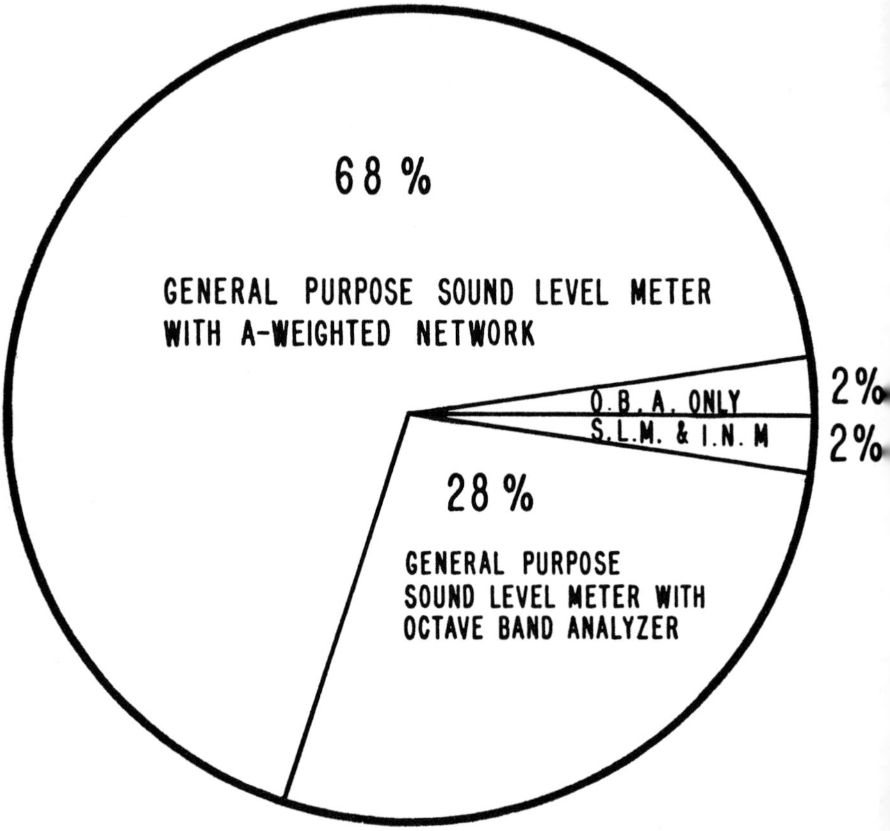

FIGURE 5.3. Instrumentation used to make noise measurements.

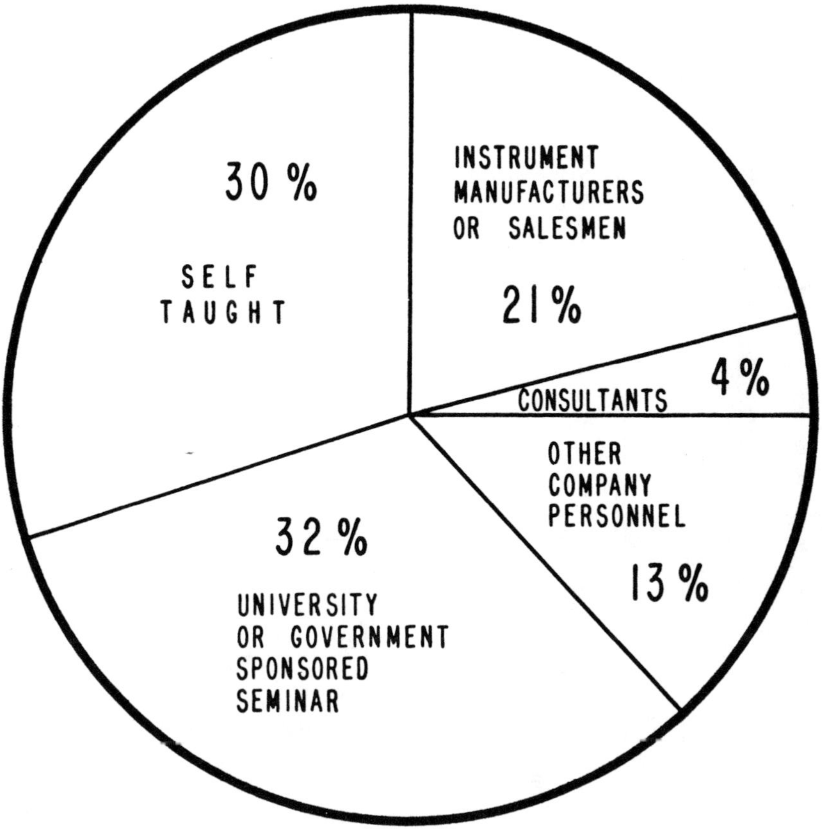

FIGURE 5.4. Training of personnel making noise measurements.

TABLE 5.8

SIX COMPANIES' PROGRAMS OF DISCIPLINARY ACTION FOR NOT WEARING EAR PROTECTORS

1	2	3	4	5	6
Oral Warning	Call to Employee's Attention	Two Verbal Warnings	Verbal Warning	Supervisor Report to Union	Verbal Warning
Letter in Personnel Folder	Notify Foreman of Infraction	Formal Written Warnings	Written Warning if within Six Months of No. 1	Union Responsible for any Disciplinary Action	Formal Letter in Personnel File
Two-Day Suspension	Verbal Warning	Transfer or Termination	Three-Day Suspension if within Six Months of No. 2		Letter to Safety Supervisor
Five-Day Suspension	Written Warning in Personnel Folder	Written Warning in Personnel Folder	Five-Day Suspension if within Six Months of No. 3		Termination of Employment
Ten-Day Suspension			Discharge if within Six Months of No. 4		

were generally the same as standard procedures used in the overall safety program (e.g., for the habitual "forgetting" of hard hats or eye protectors). Examples of disciplinary programs used in six industries are listed in Table 5.8.

In interpreting the results of this survey, it should be pointed out that each plant made its own noise level evaluations and filled in the questionaire. It is unlikely that a single standard procedure was utilized. The results of the survey do substantiate the fact that noise is indeed a common occupational hazard which affects a large number of workers.

SUMMARY

The use of hearing protectors will not provide OSHA compliance unless engineering controls are being simultaneously developed, and will not guarantee 100% prevention of all hearing losses. The use of these devices, however, supplemented with an audiometric testing program, will effectively protect a plant against thousands or even millions of dollars in payments for worker's compensation claims.

REFERENCES

1. Guignard, J. C., "A Basis for Limiting Noise Exposure for Hearing Conservation," Aerospace Medical Research Laboratory, Wright-Patterson AFB, Ohio and EPA, Washington, D.C., Joint EPA/USAF Study AMRL-TR-73-90 and EPA-550/9-73-001-A, 1973.

2. Johnson, D. L., "Prediction of NIPTS Due to Continuous Noise Exposure," Aerospace Medical Research Laboratory, Wright-Patterson AFB, Ohio

and EPA, Washington, D.C., Joint EPA/USAF Study AMRL-TR-73-91 and EPA-550/9-73-001-B, 1973.

3. Baughn, W. L., "Relation Between Daily Noise Exposure and Hearing Loss Based On the Evaluation of 6,835 Industrial Noise Exposure Cases," Joint EPA/USAF Study, prepared for 6570th Aerospace Medical Research Laboratory, Wright-Patterson AFB, Ohio, AMRL-TR-73-53, 1973.

4. Passchier-Vermeer, W., "Hearing Loss Due to Exposure to Steady-state Broadband Noise," Instituut Voor Gezondheidstechniek, Sound & Light Division, Report 35, April, 1968 (with supplement, 1969).

5. Baughn, W. L., "Noise Control - Percent of Population Protected," *Int. Audiol.*, Vol. 5, pp. 331-338, 1966.

6. Robinson, D. W., "Estimating the Risk of Hearing Loss Due to Exposure to Continuous Noise," *Occupational Hearing Loss*, British Acoustical Society Special Volume No. 1, London and New York, Academic Press, 43-62, 1971.

7. Kryter, K. D., *The Effects of Noise on Man*, Academic Press, New York, 1970.

8. Guignard, J. C., and Johnson, D. L., "A Summary of the Basis for Limiting Noise Exposure for Purposes of Hearing Conservation," presented at NOISEXPO, Chicago, Illinois, June 4-6, 1974.

9. "Information on Levels of Environmental Noise Requisite to Protect Public Health and Welfare with an Adequate Margin of Safety," U.S. Environmental Protection Agency, Washington, D.C., EPA-550/9-74-004, 1974.

10. "Impact of Noise Control at the Workplace," Bolt,

Beranek & Newman Report No. 2671, October 29, 1973.

11. Slawinski vs. J. H. Williams and Co., 273 App. Div., 826, 298, New York, 546, 81 N.E. (2nd) 93.

12. "Survey of Hearing Conservation Programs in Industry," U.S. Department of Health, Education and Welfare Publication No. (NIOSH) 75-178, 1975.

13. "Guidelines to the Department of Labor's Occupational Noise Standards," U.S. Department of Labor Bulletin 334, Revised 1971.

Chapter 6
BENEFITS OF NOISE ABATEMENT

In assessing the economics of a noise abatement program, the total economic implication of the program must consider benefits as well as costs. These benefits may be classified into three categories:

I. Benefits to the health and safety of employees.

II. Benefits relating to morale and productivity.

III. Governmental incentives and penalties.

An employer has a clear moral obligation to implement all noise abatement measures related to category I, whether or not the economic benefits exceed the costs. The protection of human health must be considered a primary goal of our society.

The creation of a pleasant working environment, beyond that necessary to protect the health and safety of workers, does not carry such a moral obligation. A more stringent viewpoint in the benefit/cost analysis must be taken for proposed programs in category II. While it is obvious that a quiet and environmentally controlled workplace would be more desirable from an employee point of view, workers have been generally compensated for their environmental exposure and heavy task requirements by increased salaries. Exhorbitant expenditures for noise control, for the sake of creating a more pleasant work environment rather than to provide safety and health protection, may seriously challenge an industry's profitability, particularly in view of increased required expenditures for other OSHA regulations such as air and water pollution controls.

Similarly, an industry must not be naive by ignoring both the direct and intangible benefits of a noise abatement program. There is an obvious trend in industry to improve the working environment. With the increased quality of life brought about in recent years, workers are becoming more reluctant to work in poor environments. Some industries are finding that they are losing top workers to nearby plants with air conditioned or sound-controlled facilities. Decreased productivity is currently a problem of national concern, and it must be recognized that decreased morale due to dissatisfaction with environmental conditions is a major contributor to the problem. It is not a coincidence that many of the largest, most productive, and most profitable industries in the country are also the quietest, and possess pleasant overall working environments.

Clearly, an economic analysis is necessary to guide management's decisions relating to noise control. Potential economic benefits to an industry due to the implementation of a noise abatement program include the following:

1. Decreased probability of payment of worker's compensation benefits for hearing loss.

2. Reduced worker's compensation insurance premiums.

3. Reduced incidences of physical and psychological disorders.

4. Improved morale.

5. Increased worker performance.

6. Reduced accident rates.

7. Minimization of absenteeism.

8. Reduced turnover rates.

9. Improved operations.

10. Energy savings.

11. Reduced penalties for OSHA citations.

12. Reduced legal fees for OSHA citation defense.

13. Loan procurement opportunities for small business.

14. Tax advantages with accelerated depreciation allowances.

HEARING LOSS

To assess the economic benefits due to prevention of hearing loss, it is necessary to distinguish hearing loss prevention achieved by engineering noise reduction from that achieved by personal protective devices. There are two strong divergent opinions regarding the relative effectiveness of the two approaches. The OSHA regulations favor engineering controls, and recognize personal protective equipment only as a temporary measure. However, there exists the opposing opinion that ear plugs and muffs do provide adequate protection. Without taking sides in this controversy, it would be agreed that the effectiveness of hearing protection in preventing hearing loss would be dependent upon:

a. Wearing of the devices.

b. Proper fitting of the devices.

c. Audiometric testing to verify effectiveness.

The dollar value of hearing loss compensation awards presently shows an upward trend, with one recent award of \$30,000. For estimating purposes, an average payment of \$2000 may be assumed. The total loss compensation liability expected for any given working environment may be estimated from:

$$C = \$2000(N) \tag{6-1}$$

where: C = cost of hearing loss compensation in dollars.

N = total population within a working environment expected to exhibit significant hearing loss.

Clearly, not every employee who incurs a permanent hearing loss will file for worker's compensation. However, an employer is legally liable for such compensation where hearing loss is incurred, whether or not it is claimed. Furthermore, an employer has the responsibility of protecting the safety and health of his employees. Thus, a direct mutual benefit to both the employee and management is realized whenever hearing loss is prevented, through either engineering controls or personal protective equipment.

THE EFFECT OF NOISE ON PERFORMANCE

The non-auditory effects of noise may be as important to industrial managers as the risk of hearing damage. Beyond the issue of legal compliance, the administrator must consider the following questions in establishing a noise control policy:

1. Are existing productivity and morale problems due

to employee dissatisfaction with an undesirable, noisy work environment?

2. Can noise control engineering expenditures be justified by anticipated productivity gains due to such a measure?

The answers to these questions involve many complex variables, and scientific research in the area of noise influence on workers' performance is not abundant. While industrial psychologists and acousticians have documented productivity results due to noise reduction, often the studies were not conducted under controlled conditions, and do not yield consistent data. While one investigator reports that noise costs American industry $4,000,000 annually in accidents and lost production[2], other studies find noise to increase worker productivity[3]. Regarding the conflicting performance data, McCormick has suggested that, "By selection of relevant studies, one can 'prove' that noise (1) produces a decrement in human performance, (2) has no effect on such performance, or (3) produces an increment in performance."[4]

Employee attitudinal patterns directly affect the following:

1. Productivity
2. Morale
3. Absenteeism
4. Turnover rate

Noise is unquestionably one of the many factors involved in forming an employee's overall attitude toward his job. Employee attitude is a function of both environmental compatibility and human interaction with management and co-workers.

PRODUCTIVITY

It is evident that when a task involves auditory signals, whether speech or nonspeech, noise at any intensity sufficient to mask or interfere with the perception of these signals will interfere with the performance of the task. When mental or motor tasks do not involve auditory signals, the effects of noise on their performance have been difficult to assess[5].

A number of general conclusions have emerged, however[1]:

1. Steady noises without special meaning do not seem to interfere with human performance unless the noise level exceeds about 90 dBA and not consistently even then[6].

2. Intermittent and impulsive noises are more disruptive than steady-state noises[7].

3. High frequency components of noise (above about 2000 Hz) usually produce more interference with performance of a task[8].

4. Noise usually does not influence the overall rate of work, but high levels of noise may increase the variability of the work rate. There may be "noise pauses" or gaps in response[9], sometimes followed by compensating increases in work rate.

5. Noise is more likely to reduce the accuracy of work than to reduce the total quantity of work[9,10].

6. Complex or mentally demanding tasks are more likely to be adversely influenced by noise than simple tasks.

7. Some tasks have been described in the literature as particularly sensitive to noise. Among them

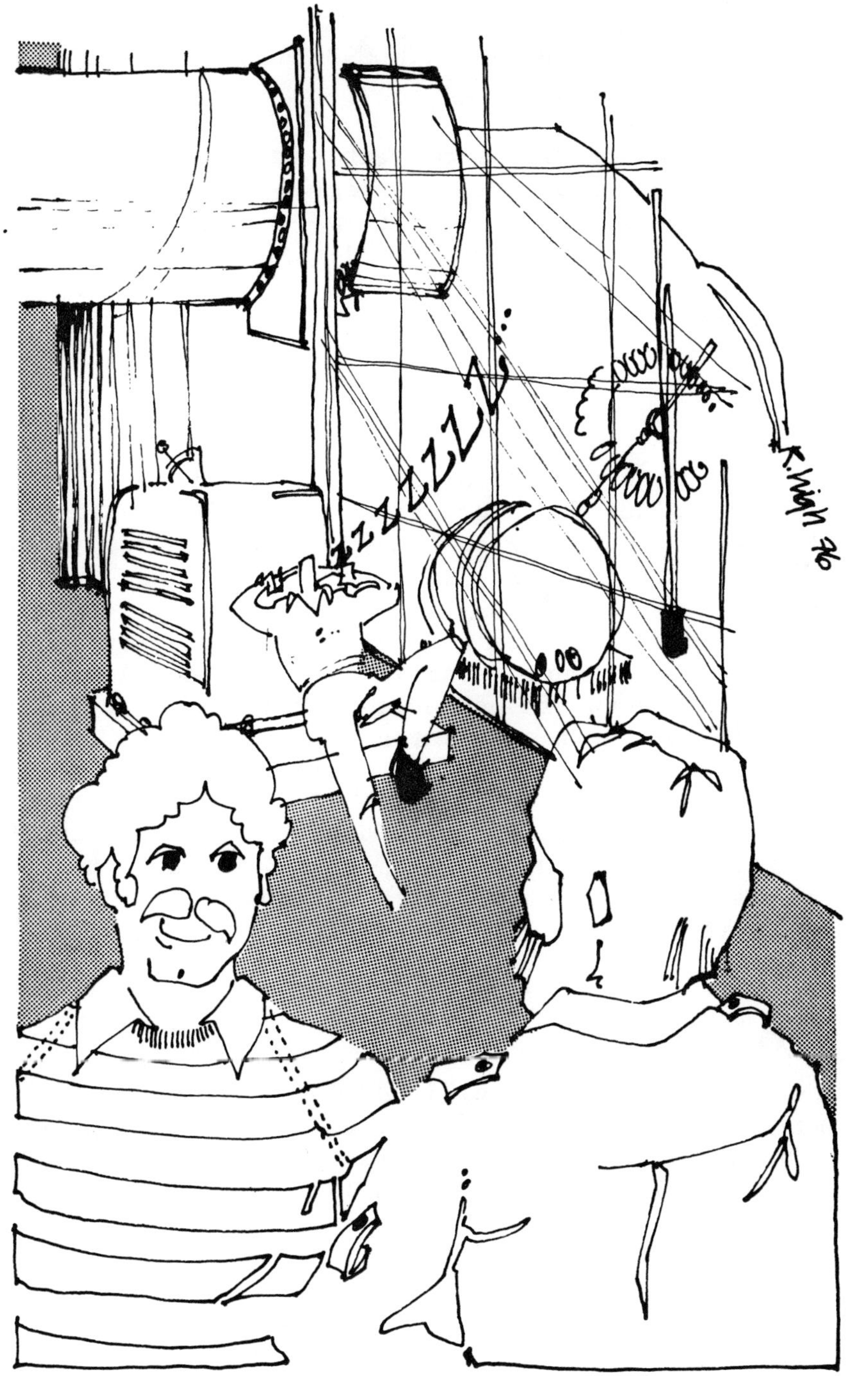

"There really has been a change in production and performance since we quieted the plant."

are tasks of vigilance, information gathering and analytical processes[12]. Vigilance activities are not repetitive, do not allow for self-pacing, and demand rapid and accurate decisions.

8. Higher order tasks requiring greater mental faculties (problem solving, creative thinking), although dependent upon the individual, are generally disrupted by a lower noise intensity.

9. An individual may maintain a constant level of performance for a primary task in the presence of noise; however, errors may result in a secondary or subsidiary task[13].

10. More moderate levels of noise may actually benefit task performance relative to quiet conditions. The presence of such noise may provide for a more uniform acoustic background, masking stray sounds which could otherwise be audible and cause distraction. On the positive side, pulsating-type sounds may pace or drive performance, and, in effect, reduce fatigue from tasks that are monotonous or boring. The rhythm component in music may serve this function also[3,14].

MORALE

Increased morale, per se, is not an economic benefit. However, its resulting effects on worker performance may be a direct economic benefit. It is difficult to quantitatively determine the relationship between occupational noise exposure and employee performance. The so-called intangible benefits related to morale defy economic analysis on a large scale basis. On an individual plant basis, however, management can review its profit, productivity, turnover, accident rates and absenteeism in comparison to industry-wide and regional averages. Where problems in these areas are recognized, management should carefully examine possible correlation with noise and other environmental conditions.

The effect of noise on performance has been discussed previously. This information should be reviewed as background in assessing possible productivity benefits in a particular plant.

ABSENTEEISM

The Department of Labor statistics have shown an increasing trend in the unscheduled absence rate in American industry in the last 10 years. These unscheduled absences, which nationally amount to 1.5% of the potential man hours available in a typical week, result in the following direct costs to industry[20]:

1. Wages paid under continuation plans.

2. Fringe benefits to absent workers (which accrue anyway).

3. Overtime pay to other workers to compensate for the loss of scheduled output per shift.

4. Padded employment to compensate for expected absences.

5. Poor quality or defective products because of absence of key personnel.

While reasons other than noise, such a boredom, heat, dust, and poor working conditions may cause increased absenteeism, recent NIOSH studies show that such absences are significantly higher in high noise areas than in low ones. The increase in overall absenteeism among workers in noisy areas as compared with non-noisy areas is on the order of 1.23 workers per hundred per day. At a $4.50 basic wage, this represents an average of $115 per worker per year.

The NIOSH study also showed that the number of lost-time accidents was greater in high noise areas. The statistics indicated that 35% of workers in noisy areas incurred 15 or more accidents in a 5 year period as compared with fewer than 5% in the same period in quieter areas. These figures must be applied with caution because of the many interrelated variables which may have a bearing on accident rates among workers.

TURNOVER RATES

Due to the complexity of evaluating an employee's overall satisfaction with his job, a quantitative relationship between noise exposure and job turnover has not been established. Except in extreme cases, noise would not be expected to be the greatest factor relating to job acceptability in comparison to other parameters such as pay, management attitudes, and overall working conditions. In some cases, annoyance due to noise may be the final stress which causes a good employee to quit. In other cases, an industry may experience no turnover problems due to noise. The author was reminded recently that these intangible, morale-related benefits vary from industry to industry. In explaining possible benefits of reducing existing 100 dBA noise levels, the benefit of an improvement in the employee turnover rate was pointed out. The plant manager responded that this plant had not had an employee quit for six years! Clearly, a decreased turnover rate would not offer a potential noise abatement benefit in this particular plant.

In the absence of any reliable existing guidelines regarding noise-related turnover, the industrial manager should be aware of the *potential* influence of noise where turnover problems are observed. In assessing the influence of noise in any turnover problem investigation, an in-depth study which considers all variables involved would be recommended.

NOISE AND ACCIDENTS

Cohen has complied a 5 year study of medical, attendance, and accident files for approximately 1000 workers in factory situations[15]. Five hundred of these workers were employed in noise levels of 95 dBA or above, and 500 in 80 dBA or below. The workers located in the presence of the higher noise levels showed significantly greater numbers of job-related accidents, sickness and absenteeism than their counterparts in the quieter jobs. The results of this study are summarized in Table 6.1[16]. However, the reader is cautioned against drawing definitive conclusions because, as Cohen pointed out, the types of jobs in the noisy and quieter areas could not be equated. For example, possibly the tasks in the noisy areas were inherently more hazardous[1].

It has been suggested[17] that noise levels may contribute to worker accidents by virtue of the masking effects of warning signals or shouts of caution. Since warning signals may be designed to generate sound levels in excess of any ambient sounds, masking effects may generally be overcome. Beyond isolated situations related to masking of otherwise detectable hazards, there appears to be no evidence establishing a causal relationship between noise levels and accidents[14]. It has also been reported that hearing impairment is not a significant contributory factor to the frequency of work-related injuries in long term employees[18].

A study conducted in a boiler making facility by Raytheon Corporation, under contract to NIOSH, produced the following results:

TABLE 6.1

COMPARISON OF TYPICAL FREQUENCIES OF OCCURRENCE OF COMPLAINTS, DISCRETE ABSENCES, AND ACCIDENTS PER WORKER IN HIGH AND LOW NOISE LEVEL AREAS OF A BOILER FABRICATION PLANT FOR A FIVE-YEAR PERIOD, 1966-1970

	Typical Frequency of Occurrence per Worker	
	Below 80 dBA	Above 95 dBA
Total Sample	449	456
Diagnosed Medical Problem	0.4	3.9
Symptomatic Complaints	0.3	1.7
Discrete Absenses*	4.2	30.3
No. of Accidents	0.4	9.0

* Number of separate absence occurrences lasting one day or more.

1. Diagnosed medical disorders were reduced from 40% to 24% of the work force studied after implementation of the hearing protection program.

2. Ten percent of the workers had up to 14 accidents subsequent to the hearing protection program, whereas sixteen percent of the workers had up to 16 accidents prior to the use of hearing protection.

3. Forty-three percent of the workers had 30 or more days of absenteeism prior to the hearing protection program. This rate was reduced to 12% after the hearing protection program was in effect.

For additional information on the effect of noise on performance, the reader is referred to six comprehensive reviews of the literature available on this subject[1,5,6,7,12,19].

NOISE CONTROL AND ENERGY CONSERVATION

It is often said that the noise energy generated by a machine is wasted energy, and that reduction of noise is therefore an energy conservation measure. It must be recognized, however, that the actual acoustic energy associated with a noisy machine is extremely small. A sound power level of 100 dBA, for example, results from an acoustic power of only .01 watt. The noise radiating efficiency of most machines ranges from .001% to 1%. A jet aircraft engine, the loudest common noise source, is only a .0001% efficient radiator of sound. Thus, the reduction of noise in a machine would not alone result in any significant energy savings.

Modification of machine or operational parameters directly associated with noise generation, however, can result in very significant energy and cost savings.

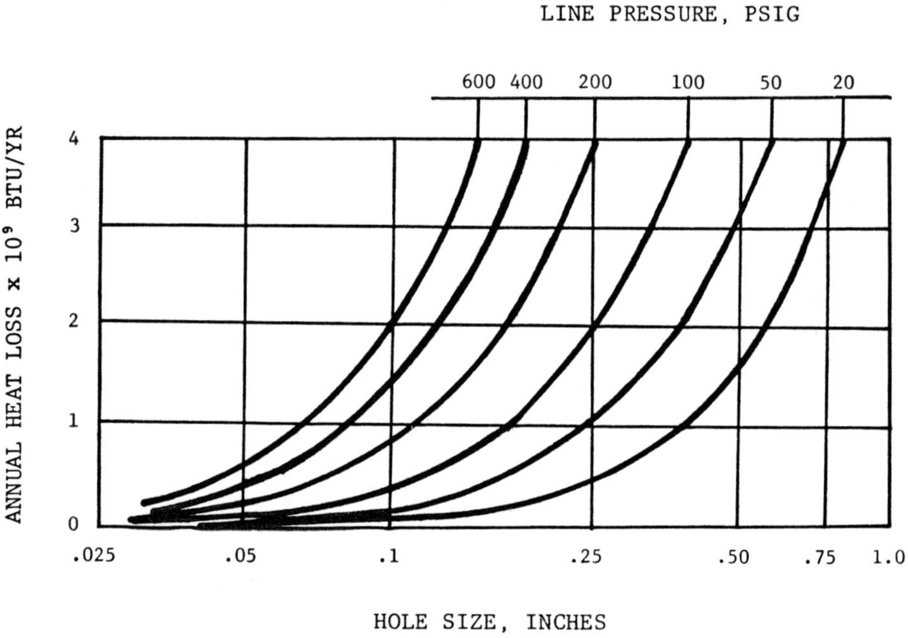

FIGURE 6.1. Annual heat loss from steam leaks. *(Adapted from NBS Handbook 115.*[20]*)*

One example is the abatement of noise associated with air discharges. Tables 6.2, 6.3, and 6.4 summarize losses for various piping systems. The cost of a small air leak or unnecessary air discharge is considerable in terms of dollars wasted over a period of time. Table 6.5 shows the cost of a continuous air leak or discharge from a sharp-edged orifice at 100 psig with an air cost rate of $0.07 per 1000 cubic feet.

Similarly, steam leaks generate excessive noise levels and waste energy. Figure 6.1 illustrates the annual heat loss from a steam leak. The installation of a steam trap will allow the passage of condensate, air and non-condensable gas from the steam piping and equipment while preventing the loss of steam. In many plants, tubing is connected directly to the steam piping in lieu of the use of steam traps. This obsolete practice can result in both excessive energy costs and OSHA noise citations.

In other cases, sound generation indicates a machine malfunction, which wastes energy. One example would be vibratory scrap conveyors which draw extremely low current when properly adjusted. The conveyors will function when not optimally tuned, but if maintenance is performed for the purpose of noise control, an energy and cost savings may be realized.

Often the consideration of replacement of older equipment with newer generation, quieter equipment can be cost-justified in terms of energy savings. In one plant, the replacement of a noisy pump with a new, quiet rotary pump paid for itself within weeks due to reduced energy costs. The replacement of fan blades with aerodynamically efficient blades can result in both energy and noise reductions.

TABLE 6.2

NATURAL GAS LOSS PER YEAR FROM LEAKS IN UNDERGROUND PIPELINES AT VARIOUS PRESSURES (KSCF)
(Adapted from NBS Handbook 115)

Hole Dia., in.	Line Pressure (psig)						
	0.25	5	25	60	100	300	500
1/64	1	4	10	20	30	80	140
1/32	2	6	20	35	60	160	250
1/16	10	40	100	200	320	900	1,500
1/8	50	200	600	1,200	1,800	5,000	8,200
1/4	250	1,200	3,300	6,500	10,300	28,300	46,500
1/2	1,400	6,600	18,800	37,300	58,000	156,500	263,000

TABLE 6.3

NATURAL GAS LOSS PER YEAR FROM LEAKS IN ABOVE GROUND PIPELINES AT VARIOUS PRESSURES (KSCF)
(Adapted from NBS Handbook 115)

Hole Dia., in.	Line Pressure (psig)						
	0.25	5	25	60	100	300	500
1/64	5	26	69	136	212	581	953
1/32	21	102	277	544	846	2,330	3,810
1/16	85	409	1,110	2,180	3,390	9,300	15,300
1/8	341	1,640	4,430	8,700	13,500	37,200	61,000
1/4	1,360	6,540	17,700	34,800	54,200	149,000	244,000
1/2	5,450	26,200	70,900	139,000	217,000	595,000	977,000

TABLE 6.4

ANNUAL HEAT LOSS FROM COMPRESSED AIR LEAKS
(Adapted from NBS Handbook 115)

Hole Diameter, in	Free Air Wasted, cu ft/hr, by Air Leak at*:	Fuel Wasted MBTU/yr**	Cost of Power Wasted, $/yr at Unit Power Cost of***:		
			$0.010/kWh	$0.015/kWh	$0.020/kWh
100 psig					
3/8	79,900,000	2190	2190.00	3280.00	4370.00
1/4	35,500,000	972	972.00	1460.00	1940.00
1/8	8,880,000	243	243.00	364.00	486.00
1/16	2,220,000	60.6	60.70	91.00	121.00
1/32	553,000	15.1	15.10	22.70	30.30
70 psig					
3/8	59,100,000	1320	1320.00	1980.00	2650.00
1/4	26,200,000	587	587.00	881.00	1170.00
1/8	6,560,000	147	147.00	220.00	294.00
1/16	1,640,000	36.6	36.60	54.90	73.30
1/32	410,000	9.2	9.18	13.80	18.40

* Based on nozzle coefficient of 0.65.
** Based on 10,000 BTU fuel/kWh.
*** Based on 22 brake HP/100 cu ft free air/min for 100 psig air and 18 brake HP/100 cu ft free air/min for 70 psig air.

TABLE 6.5

MONTHLY COST AND ENERGY CONSUMPTION OF AN AIR LEAK[21]

Diameter of Opening (Inches)	Cubic Feet Wasted Per Month	Cost of Waste Per Month
1/32	45,000	$3.18
1/16	182,000	$12.75
1/8	729,000	$51.00
1/4	2,920,000	$204.00

Note: Leakage is assumed to be as from a sharp-edged orifice continuously at 100 psig with air at 7¢ per 100 cu.ft.

MAINTENANCE

Maintenance problems can also result in noise problems. A case in point was an OSHA noise citation due to excessive noise from a gear drive. An inspection of the gear drive revealed that the bearings were bad. After the bearings were replaced, the gear drive noise level was reduced, and compliance was achieved.

Machine inefficiencies, wear and malfunction can result in significant increases in noise levels. The goal of normal plant maintenance is to keep machinery in proper condition for efficient production. If noise criteria are not assigned a level of priority, a machine with noise-producing maintenance problems, but which is still operating at 100% efficiency, would not justify maintenance attention. The control of noise by means of operational maintenance must be viewed as an added discipline and an added responsibility to normal maintenance procedures. Noise control through maintenance would require an educational program for maintenance personnel both to create "noise awareness" and to provide them with the technical background necessary to cope with the unique engineering features associated with noise reduction. Additional instrumentation may be required to fulfill the added responsibility.

The most effective program would involve creating a "noise awareness" among all personnel, and designating individuals whose primary duties are those related to plant-wide noise reduction. These duties might also include general energy conservation.

A good preventive maintenance program has several benefits in addition to noise reduction:

1. Machinery breakdowns are minimized.

2. Machine repair costs are lowered.

3. Time is saved by the maintenance staff.

4. Production schedules may be controlled better.

5. Energy costs are reduced.

Dynamic balancing of machinery is an important element in a successful maintenance program. Again, balancing costs of rotating machinery such as fans are generally more than offset in terms of minimization of equipment failure.

GOVERNMENTAL INCENTIVES

A direct economic benefit for noise control expenditures may be derived from tax deductions allowed for environmental improvements. Accelerated depreciation schedules may be allowed for certain environmental-related projects. IRS Codes and Regulations, Sections 57 and 169 outline these benefits.

Financial assistance, otherwise unobtainable, may be available for OSHA compliance programs. Section 28 of the Occupational Safety and Health Act authorizes the Small Business Administration (SBA) to make loans to assist small businesses to meet the standards under Section 6 of the Occupational Safety and Health Act or under state laws. Such help is made available if SBA determines that the business is likely to suffer substantial economic injury without assistance. OSHA has worked out arrangements with SBA to aid in the making of such loans in appropriate cases.

OSHA bulletin 2005 outlines the following stipulations regarding what the loan may be used for:

1. An employer may use the proceeds to construct a new building, even in a new location, to replace an old

building where remodeling is not feasible or to replace rented quarters when needed upgrading arrangements cannot be arranged. Any upgrading in size or quality may not exceed corresponding criteria under the SBA's Displaced Business Loan (DBL) program.

2. Bank loans may be refunded when the terms of such loans (intended for permanent financing) prove too burdensome for the borrower and the bank itself requests refunding because it realizes a longer term will benefit the borrower. The loan may not be used to effect a take-out from a possible loss. Bank loans also may be refunded when they were rushed through for short term improvements to meet compliance requirements.

3. Working capital may be provided to replace working capital expended for compliance (such as meeting construction time limits). It also may be used, when construction is involved and operations are curtailed, to meet continuing fixed costs such as payment on equipment notes and mortgages to help finance start-up costs, and to finance operating changes required to come into compliance.

4. Tenants may obtain loans to finance the purchase of equipment or upgrade the plant.

When an employer requests a loan to bring his establishment into compliance, he must submit to SBA:

a. A statement of the conditions to be corrected.

b. A reference to the OSHA standards that require correction.

c. A statement of his financial condition that necessitates applying for a loan.

Specifically, the employer should do the following:

1. He should obtain from a licensed professional engineer and/or architect a report on the work to be done. This report should cite existing conditions, the standards that require the work, and plans and specifications of sufficient clarity to permit OSHA to determine that the work will bring the establishment into compliance.

2. He should submit to SBA a copy of this report, along with any background material.

If the industry has already been cited, the applicant must also furnish SBA a copy of the OSHA citation(s).

SUMMARY

The benefits of a noise reduction program may include a wide variety of items related to health and safety, productivity, and governmental incentives. While each of the potential benefits outlined in this chapter may not be applicable to all noise control projects, it is important that all possible benefits be assessed in evaluating proposed noise projects. Often proposed noise abatement programs are unnecessarily postponed or rejected due to cost considerations because management has viewed only project costs but has failed to recognize the more subtle but direct economic benefits of the program.

REFERENCES

1. "Public Health and Welfare Criteria for Noise," U.S. Environmental Protection Agency, 550/9-73-002, July 27, 1973.

2. Schweisheimer, W., "Noise - Another Pollution," *Australian Printer*, July 1971, pp. 28-29.

3. Wokown, W., "Effects of Music on Work Performance," Technical Memorandum 1-68, U.S. Army Human Engineering Laboratory, Aberdeen Proving Ground, Maryland, 1968.

4. McCormick, E. J., *Human Factors Engineering*, McGraw-Hill, New York, 1970.

5. Kryter, K. D., *Effects of Noise on Man*, Chapter 13, Academic Press, New York, 1970.

6. Broadbent, D. E., "Effects of Noise on Behavior," *Handbook of Noise Control* (C. M. Harris, Ed.), McGraw-Hill, New York, 1957.

7. Cohen, A., "Effects of Noise on Psychological State," *Noise as a Public Health Hazard, ASHA Reports #4*, American Speech and Hearing Association, Washington, D.C., 1969.

8. Sanders, A. F., "The Influence of Noise on Two Discrimination Tasks," *Ergonomics 4*, 235, 1961.

9. Carpenter, A., "Effects of Noise on Performance and Productivity, *Control of Noise*, Symposium No. 12, 297-310, H. M. Stationary Office, London, 1962.

10. Fornwalt, N. E., "Investigation into the Effect of Intermittent Noise of Constant Periodicity vs. Random Periodicity on the Performance of an Indus-

trial Task," Master's Thesis, Dept. of Industrial Engineering, Texas Tech. College, 1965.

11. Boggs, D. H., and Simon, J. R. "Differential Effect of Noise on Tasks of Varying Complexity," *Journal of Applied Psychology*, Vol. 52, pp. 148-152, 1968.

12. Guignard, J. C., and King, P. F., "Aerodynamic Aspects of Vibration and Noise," AGARDograph #151, NATO Advisory Group for Aerospace Research and Development, Neuilly sur Seine, France, 1972.

13. Finlelman, J. M., "Effects of Noise on Human Performance," *Sound and Vibration*, Vol. 10, No. 9, September 1975, pp. 26-28.

14. "Criteria for a Recommended Standard: Occupational Exposure to Noise," NIOSH, 1972.

15. Cohen, A., "Industrial Noise and Medical Absense, and Accident Record Data on Exposed Workers," presented at International Congress on Noise as a Public Problem, Dobrovnik, Yugoslavia, 1973.

16. Cohen, A., "Extra-Auditory Effects of Occupational Noise: Effects on Work Performance," *National Safety News*, September 1973, pp. 68-76.

17. Wilson, A., *Noise*, Her Majesty's Stationery Office, London, 1963.

18. Straub, C. S., "Hearing Loss and Accidents," *Sound and Vibration*, Vol. 8, No. 1, January 1974, pp. 22-23.

19. Gulian, E., "Psychological Consequences of Exposure to Noise, Facts and Explanations," presented at International Congress on Noise as a Public Health Problem, Dobrovnik, Yugoslavia, 1973.

20. Gatts, R. R., Massey, R. G., and Robertson, J. C., "Energy Conservation Program Guide for Industry and Commerce," *National Bureau of Standards Handbook 115*, 1974.

21. Gibbs, C. W. (ed.), *Compressed Air and Gas Data*, Ingersoll-Rand, 1971.

Chapter 7
WHAT WILL IT COST ?

In Chapter 6, benefits which result from noise abatement were discussed. These benefits represent a potential cost savings. In this chapter the costs for noise abatement are examined.

When noise control measures are "added on" to an existing facility, several first costs should be considered:

1. Basic material and installation costs.

2. Production downtime as a result of the noise abatement installation.

In addition, the noise abatement measures may increase production and maintenance costs.

When noise control measures are implemented in the design stage of new facilities or in expansion projects, options of equipment and employee location are available to the designer to minimize noise problems. In this case the noise abatement cost is the increase in cost of the materials and installation over the cost of the same system without noise control.

The increased cost for quieter technology in new equipment is found to vary. Often noise control features are included as standard items in new equipment. In other cases, noise control features may add 10% or more to the cost of new equipment. The total cost of noise abatement measures, or for achieving OSHA compliance, is generally significantly greater when implementation occurs on existing machinery rather than in initial design.

NOISE CONTROL MATERIAL AND SYSTEM COSTS

A recent study of industry-wide noise abatement costs was sponsored by the U.S. Department of Labor to establish the impact of the OSHA noise control regulations[1]. A summary of the cost estimates utilized in this study is presented in Table 7.1. A compilation of typical noise control costs based on manufacturer's information[2,3], is presented in Table 7.2.

Material costs for mufflers or silencers can be estimated on a dollar per square foot basis for the common lengths available, as shown in Table 7.3. Included in this table is a breakdown of muffler costs, differentiating between heavy gage industrial types and lighter gage types used in HVAC systems.[2]

Typical costs of standard and quiet motors are presented in Table 7.4.

INSTALLATION COSTS

Installation costs for noise control materials can vary from about 50% to 200% of material costs, depending upon the need for structural support, the local labor union requirements, and the overall complexity of the installation. Some installation cost savings could be expected for new installations over noise control retrofit situations.[2] Typical installation costs for barriers are given in Table 7.5.

TABLE 7.1

ESTIMATED INSTALLED COSTS
OF NOISE CONTROL SYSTEMS[1]

Equipment	Estimated Cost*
Machine Enclosures	
Low (screw machine)	$1,500
High (special press)	$40,000
Partitions	
Typical	$8/sq ft
Mufflers	
Small Pneumatic	$10
500 HP Diesel Engine	$100-$200
Industrial Ventilating Fans	$50-$200
Exterior Exhaust	$5,000-$10,000
Motor Enclosures and Mufflers	$1,000-$4,000
Twisting Machine Drive	$50/spindle
Front-loaders; Fan, Muffler, Partial Enclosure	$200-$500
Building Absorptive Treatment	$1.50-$2.00/sq ft
Personnel Enclosures	
Range	$3,000-$15,000
Average	$8,000
Air Coolers	$4,000
Compressors	
Centrifugal 5000 HP+	$10,000
Less than 5000 HP	$6,000

TABLE 7.1 (Continued)

Equipment	Estimated Cost*
Reciprocating	$8,500
Electric Motors	
25-100 HP	$1,000
100-200 HP	$2,000
Large Motors Requiring Total Enclosure	$4,000
Heaters	$250/MMBTU/hr
Valves: Special Low Noise	$1,000-$2,000
Piping	
4"	$4/ft
6"	$8/ft
8"	$16/ft

* Costs per unit unless otherwise noted.

TABLE 7.2

ACOUSTICAL MATERIAL COST SUMMARY

Material	Cost ($/sq ft)
4" Acoustical Panel	5.00-10.00
2" Acoustical Panel	3.00-6.00
Lead Sheet, 1 psf	.40-.75
Lead-Vinyl Sheet, 1 psf	1.70-3.00
Barium Sulphate-Vinyl Sheet, 1 psf	1.20-1.50
Vibration Damping	.25-6.00
Lead-Vinyl with 1" Foam	2.70-4.00
1" Acoustical Foam	1.00
1" Foam with Adhesive	1.25-1.50
1" Foam with 1 Mil Plastic Facing	1.25-1.50
Spray-On Ceiling Material, 1"	.35-1.00
Glass Fiber	.10-.30
1/4" Plywood	.25
3/4" Plywood	.45
18 GA Cold Rolled Steel	.30-.35
24 GA Cold Rolled Steel	.15-.20
1/4" Plexiglas	2.50
1/4" Lexan	4.75

TABLE 7.3

MUFFLER COST SUMMARY[2]

Length (ft)	Cost ($/ft^2)
LIGHT GAGE HVAC SYSTEM	
3	25-75
5	30-100
7	45-125
HEAVY GAGE INDUSTRIAL (AMBIENT AIR APPLICATION)	
3	40-75
5	55-125
7	70-150
10	100-200
HEAVY GAGE INDUSTRIAL (HOT GAS APPLICATION)	
3	125
5	200
7	250
10	350

TABLE 7.4

TYPICAL DISCOUNTED MOTOR PRICES

Motor HP	Standard Motor ($)		Quiet Line Motor ($)*	
	3600 rpm	1800 rpm	3600 rpm	1800 rpm
1½	107	117	112	103
2	125	128	124	108
3	147	143	143	123
5	183	167	167	167
7½	198	198	222	222
10	230	239	267	277
15	326	319	338	413
20	415	400	461	507
25	510	486	593	635
30	603	573	715	743
40	800	764	932	978
50	1060	972	1178	1148
60	1429	1430	1478	1509
75	1843	1817	1949	1819
100	2486	2282	2577	2314
125	3144	2945	3248	2998
150	3801	3352	3918	3478
200	5172	4316	5508	4646

* Quiet line motors are standard motors which have been modified by the equipment manufacturer in order to reduce noise.

Note: Standard motor is based on mill and chemical motor, class F insulation and a 1.15 service factor. Quiet line motor is based on class F insulation with a 1.0 service factor. The above prices include a 42% discount for standard motors and a 10% discount for quiet line motors. Prices as of December, 1974.

TABLE 7.5

CONSTRUCTION COSTS OF BARRIERS[4]

Barrier Type	Thickness	Cost/Lin. Ft. for 10-Ft. Barrier
Wooden Wall	3/4"	$12*
Earth Berm	60"(top)	$25
Concrete Wall (pre-cast cellular)	6"	$42-$55*
Brick	-	$25**
Aluminum	3"	$40*
Gabion (stone)	36"	$60*

** Estimate - Ontario Ministry of Transportation.
* Estimate - Greater London County Council.

INFLUENCE ON PRODUCTION

The implementation of a noise control system can have either a positive or negative influence upon productivity and production costs. In some cases, noise exposure problems may initiate or warrant production modification to automated or semi-automated operations, with substantial long term cost savings. In these cases the benefit of noise exposure reduction may be exceeded by productivity benefits. In other instances reduction of noise levels may be an unplanned, secondary benefit of an automation program. Productivity benefits may also be realized by a noise reduction program in terms of improved morale and reduction of physiological task interference.

When noise control systems are "added on" to an existing workable system, productivity may be adversely affected. Such influence may result from:

1. Restriction of mobility of the operator.

2. Reduction of machine capacity.

3. Reduction of visibility.

4. Remote location of the operator.

5. Increased machinc failure.

6. Increased maintenance requirements.

7. Personnel retraining requirements for modified operations.

8. Psychological resistance to change.

9. Elimination of sounds utilized by the operator to judge machine performance.

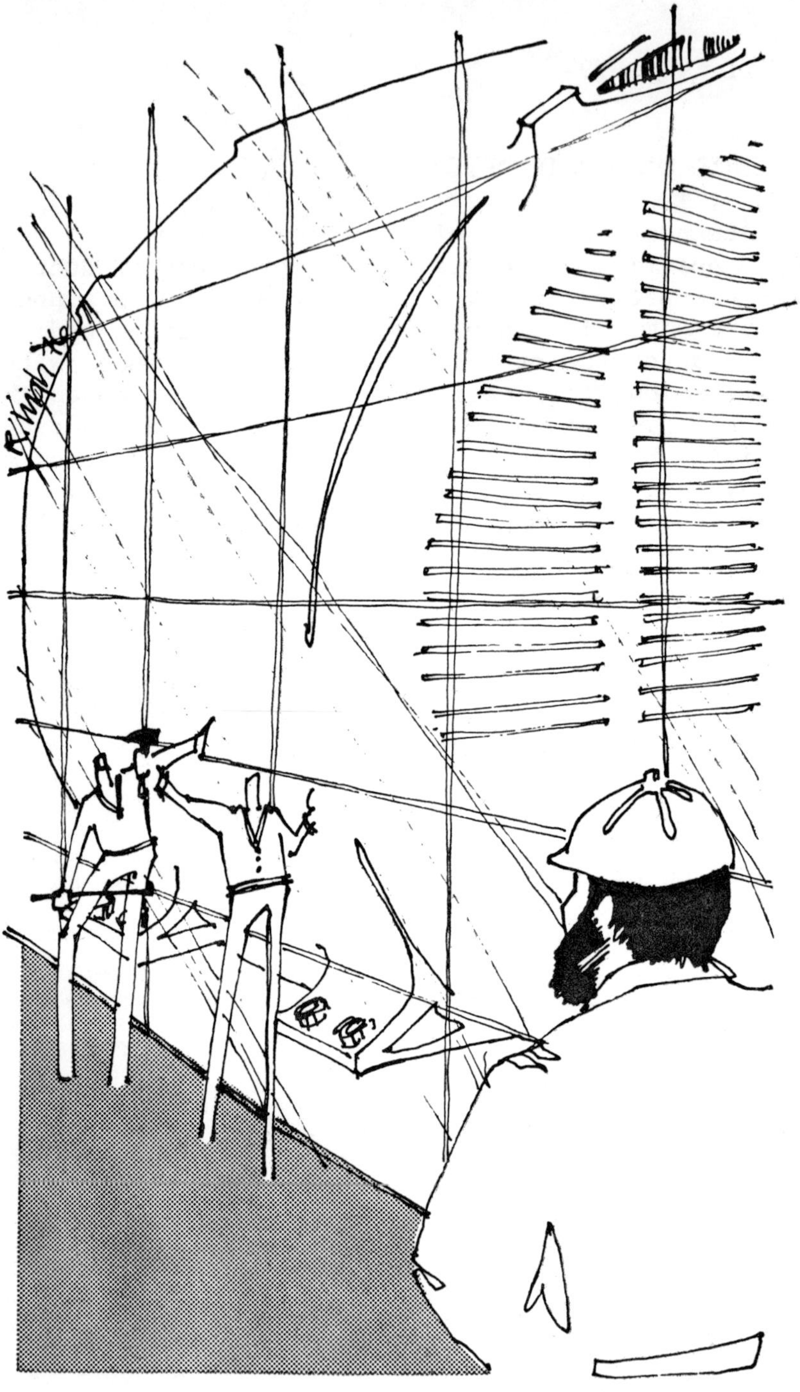

"It sure is quiet with this new noise enclosure—it's too bad we can't get in to run the machine."

TABLE 7.6

PRODUCTION PENALTIES DUE TO VARIOUS NOISE CONTROL MEASURES[5]

Control Option	Productivity Penalty
Absorption	None
Damping	None
Barriers	Up to 15%
"Glove Box" Booths	Up to 20%
Machine Enclosures	Up to 25%
Worker Shelters	None

TABLE 7.7

TOTAL EXPENDITURES FOR SAFETY AND HEALTH[6]

	Expenditure (Billions of Dollars)		
	1972	1973	1976 (est.)
All Business	2.51	3.16	3.56
Manufacturers	0.94	1.26	1.83

TABLE 7.8

SAFETY AND HEALTH EXPENDITURES
BASED ON INDUSTRY SIZE

Number of Employees	Cost ($)
Less than 100	33,000
100-500	104,000
500-1000	212,000
1000-2000	372,000
2000-5000	863,000
Above 5000	1-7 million

An estimate of productivity penalties due to various noise control measures[5] is presented in Table 7.6.

INDUSTRY WIDE COSTS

Compliance with the OSHA standards is, without a doubt, expensive. A McGraw-Hill survey in 1973 compiled the actual expenditures shown in Table 7.7 related to employee safety and health[6]. These costs represent approximately 3% of all capital spending.

In another survey, the National Association of Manufacturers queried 1150 member companies about OSHA. The NAM survey shows the cost of complying with OSHA standards as indicated in Table 7.8[6].

Although the cost of compliance with OSHA standards and the increasing costs of employee safety and health programs are clearly substantial, these costs are dwarfed by estimates of the annual cost of accidents to industry. According to the National Safety Council's latest figures, accidents cost industry $10-20 billion annually.

As assistant Secretary of Labor for OSHA, John Stender, explained to a meeting of the Associated General Contractors, "....I would like to point out that it is to your advantage economically to ccoperate with us in making your jobsite safer....Each one of you knows what it means to have a man injured on the job. In terms of time wasted, delays in meeting construction due dates, and lowered employee morale, accidents are expensive."

A recent study completed by the consulting firm, Bolt, Beranek & Newman, Inc., for the Occupational Safety and Health Administration estimated an industry-wide total cost of $13.5 billion to achieve compliance with the existing 90 dBA noise standard, and $31.6 billion for

TABLE 7.9

ESTIMATED OSHA COMPLIANCE COST FOR NOISE CONTROL (90 dBA)[1]

Industry	Cost (Millions of $)
Utilities	3,200
Nonelectrical Machinery	1,400
Fabricated Metal Products	1,100
Transportation Equipment	1,100
Textile Mill Products	1,100
Food & Kindred Products	590
Electrical Machinery	780
Primary Metals	900
Chemicals & Allied Products	1,100
Printing & Publishing	870
Lumber & Wood Products	160
Furniture & Fixtures	190
Stone, Clay & Glass	290
Paper & Allied Products	140
Rubber & Plastics Products	300
Petroleum & Coal Products	210
Tobacco	45
Apparel & Related Products	0
Leather & Leather Products	0

an 85 dBA standard[1]. These costs were estimated by industry type, as shown in Table 7.9. In July 1976 an economic impact study dealing with the proposed federal noise exposure regulation was prepared by Bolt, Beranek & Newman and published by the Occupational Safety and Health Administration. This report concluded that industry will have to spend $10.5 billion to comply with engineering requirements of a 90 dBA standard and $18.5 billion to meet an 85 dBA standard.

Not surprisingly, the U.S. Environmental Protection Agency, in support of the 85 dBA standard, believes these estimates to be high[7], while studies supported by industry indicate the actual cost would exceed the OSHA estimates. Despite this controversy, the cost of $10.5 billion remains the most recognized figure and provides an order-of-magnitude estimate with regard to the commitment that industry must make to achieve a sound controlled environment. One estimate of the actual commitment of public and private funds to noise control during the period of 1971-1980 was $2.1 billion[8].

REFERENCES

1. Bolt, Beranek & Newman, Inc., "Impact of Noise Control at the Workplace," Report No. 2671, October 29, 1973.

2. Carlson, J. P., "Cost of Noise Control as Seen by a Public Utility," presented at the 89th meeting of the Acoustical Society of America, Austin, Texas, April 8-11, 1975.

3. Dunn, R. L., "Noise Control Products," *Plant Engineering*, Vol. 25, No. 25, December 14, 1972.

4. Bragdon, C. R., "Urban Planning and Noise Control," *Sound and Vibration*, Vol. 7, No. 5, May 1973.

5. Yerges, L. F., "Methods of Noise Control for Machinery Already Installed," proceedings of NOISEXPO, September 1973, pp. 157-159.

6. Anon., "What Does It Cost to Comply with OSHA," *Occupational Hazards*, October 1973, pp. 114-117.

7. *Federal Register*, Vol. 39, No. 244, pp. 43802-43809, December 18, 1974.

8. Estimate by Michael P. Milane, Datatronic Systems Corporation, *Noise Control Report*, Vol. 2, No. 5, March 5, 1973, p. 41.

9. Warnaka, G. E., "Noise Control and its Economic Implications," presented at 89th meeting of the Acoustical Society of America, Austin, Texas, April 8-11, 1975.

Chapter 8

ESTABLISHING NOISE ABATEMENT PRIORITIES

The priority level assigned to noise reduction projects must be weighed against all other priorities of a company. Rank ordering of these priorities is dependent upon the specific demands placed on each company.

Every business has two-way relationships with groups which economists call *contributor-claimants*. Each of these groups has separate, and generally conflicting interests. Six contributor-claimants whose demands relate to noise control and their associated demands are:

Contributor-Claimant	*Demand*
1. Employees	Safety and Health Protection
2. Share Owners	Profit
3. Customers	Product Quality
4. Government	Legal Compliance
5. Community	Environmental Compatibility
6. Suppliers	Sales to Business

This relationship is shown in Figure 8.1.

A business will serve each claimant in proportion to the relative magnitude of his demand. Priorities will vary from business to business, depending upon the influence of the respective contributor-claimants. Thus, no universal rank-ordering system can be developed for establishing noise abatement priorities. Each business must recognize its objectives and demands before its

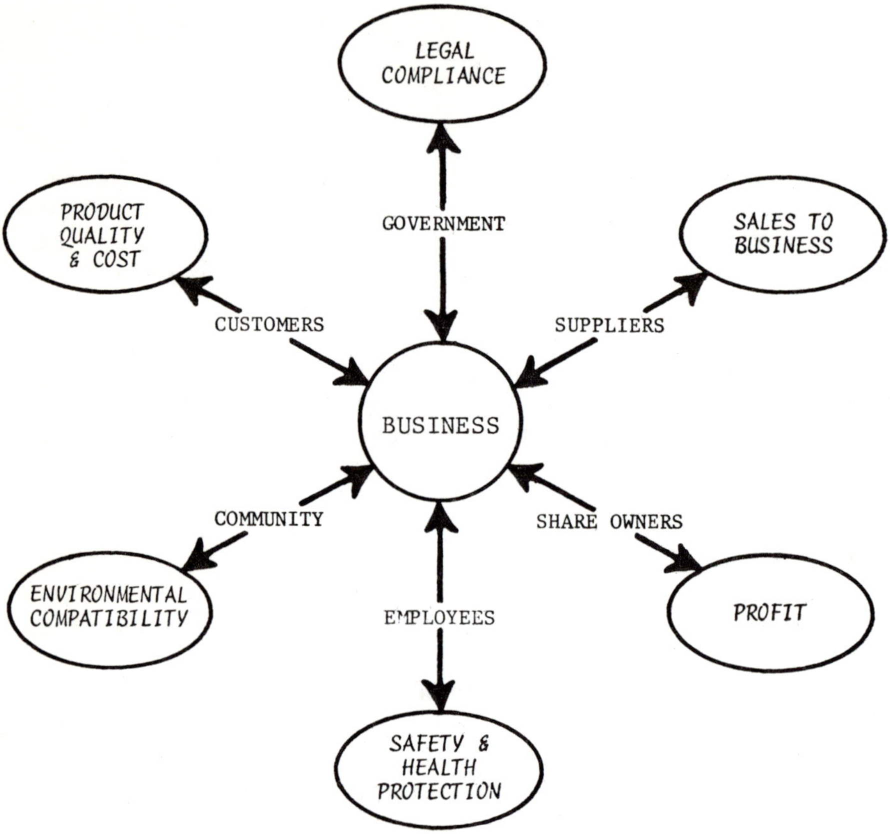

FIGURE 8.1. Responsibilities of an industry directed toward six contributor-claimants.

rank-ordering system can be established.

Three primary criteria exist for judging the relative merits of specific noise reduction techniques:

1. *Cost-Effectiveness*: The reduction in employee noise exposure as a function of quieting cost.

2. *Cost-Benefit*: An analysis of the economic factors associated with noise reduction.

3. *Equitability*: The degree to which the beneficiaries of a quieting program bear the expense of that program.

RANK ORDER SYSTEMS BASED ON SEVERITY, COST AND EFFECTIVENESS

The implementation of a noise control program requires, first, an assessment of the total in-plant noise environment. The noise of each operation should be assessed with regard to legal compliance and hearing loss potential. Following the identification of potential noise problem areas, priorities for noise abatement must be established.

Priorities established by severity of impact are based on:

- Noise intensity
- Noise exposure time
- Number of employees
- Legal requirements

The noise intensity and exposure time may be equated to a noise dose value, D, as defined in Chapter 2 (see Equation 2-1). Legal requirements may be observed by considering only noise exposures which exceed the OSHA guidelines, or noise doses which exceed unity.

A noise severity factor, *S*, may be assigned to each operation:

$$S = \sum D_i \tag{8-1}$$

where: D_i = noise dose of each employee exposed to any operation.

The effectiveness of a noise reduction measure may be evaluated in terms of the noise dose reduction for the employees affected. An effectiveness factor, *E*, may be assigned to each potential noise reduction measure:

$$E = \sum \Delta D_i \tag{8-2}$$

where: ΔD_i = reduction in noise dose of each employee due to a noise reduction measure.

The effectiveness-cost ratio of a noise reduction measure may be computed by dividing the effectiveness factor, *E*, by the cost, *C*, of noise abatement:

$$E/C = \frac{\sum \Delta D_i}{C} \tag{8-3}$$

The *E/C* ratio is a measure of hearing risk reduction per unit capital investment. Thus, for the same capital investment, the alternative solution with the highest *E/C* ratio would be the better project.

COST-BENEFIT ANALYSIS

A sophisticated cost-benefit analysis for a noise control program is usually not warranted, since is is difficult to accurately evaluate the dollar savings and costs due to noise reduction. Estimates as to the yearly cost savings due to noise reduction can be made for the following:

1. Reduction of energy costs, such as elimination of steam and air leaks.

2. Reduction of production "downtime" because of preventive maintenance.

3. Increase in production because of use of modern materials handling and automated systems.

4. Increase in employee performance because of better working conditions.

5. Reduction in risk of increased insurance premiums because of worker's compensation claims.

6. Decrease in worker's compensation claims as a result of fewer cases of noise-induced hearing loss.

The reason, of course, that noise reduction programs cannot be considered on a cost analysis basis alone is the fact that excessive noise exposure causes hearing damage. The more employees exposed to high noise levels, the greater the risk. Therefore, along with the cost-benefit analysis, a measure of hearing damage risk must be given.

RATE OF RETURN

Proposed noise reduction measures require the investment of capital for their implementation. By investing today in noise reduction, annual operating dollars over the life of the investment will be saved. A dollar in hand now can be considered more valuable than one to be received at a time in the future. Thus, a *time value* is placed on all cash flow into and out of the company.

The rate of return before taxes is a simple indicator of the relative merit of an investment. The rate of return, *i*, is found by first computing the capital recovery factor, *C*, which is the ratio of the net operating savings, *R*, to capital investment, *P*:

$$C = \frac{R}{P} \tag{8-4}$$

where: C = capital recovery factor

R = uniform series of deposits (occurs at the end of the interest period), including annual maintenance and operating savings

P = present worth (occurs at the beginning of the interest period)

The interest rate, or rate of return, *i*, is related to the capital recovery factor by:

$$C = \frac{i(1 + i)^n}{(1 + i)^n - 1} \tag{8-5}$$

where: C = capital recovery factor

i = desired rate of return or interest rate

n = number of periods that the interest is computed

The rate of return may also be found from Table 8.1.

TABLE 8.1

CAPITAL RECOVERY FACTOR AT SPECIFIED RATE OF RETURN AND PERIOD

PERIOD	8%	10%	12%	15%	20%	25%	30%	40%	50%
5	0.25046	0.26380	0.27741	0.29832	0.33438	0.37185	0.41058	0.49136	0.57583
10	0.14903	0.16275	0.17698	0.19925	0.23852	0.28007	0.32346	0.41432	0.50882
15	0.11683	0.13147	0.14682	0.17102	0.21388	0.25912	0.30598	0.40259	0.50114
20	0.10185	0.11746	0.13388	0.15976	0.20536	0.25292	0.30159	0.40048	0.50015
25	0.09368	0.11017	0.12750	0.15470	0.20212	0.25095	0.30043	0.40009	-
30	0.08883	0.10608	0.12414	0.15230	0.20085	0.25031	0.30011	-	-

EXAMPLE

A capital expenditure of \$50,000 reduces the noise from steam leaks to acceptable values. The resulting energy savings is \$10,000 per year. What is the before tax return on investment. Assume a plant life of 15 years.

Answer:

From equation (8-4),

$$C = \frac{10,000}{50,000} = .2$$

From Table 8.1, the rate of return which corresponds to C of .2 for a plant life, n, of 15 is:

$$i = 20\%$$

LEGAL INFLUENCE

It is not suggested that a company should adopt a policy which ignores legislative noise guidelines; however, other important priorities, including compliance with regulations not associated with noise must also be considered. The fact that sound levels at the present time are not universally below 90 dBA in American industry indicates that certain legal compromises have been made.

A number of legal considerations must be taken into account, and the following questions should be considered (these are applicable to all OSHA and pollution control projects)[1]:

1. When was the legislation passed?

2. What court decisions have been reached on it?

3. What trends in interpretation do the court decisions reflect?

4. If there are no court decisions, what is the meaning of the law in your attorney's opinion?

5. What is the schedule for implementation?

6. What schedule delays or revisions in implementation of the regulations have been granted?

7. What are the precise penalties for violation?

8. What do the penalties cost in dollars, other tangibles, and intangibles?

9. What are the chances of getting cited?

10. From the answers to these questions asked about each of the laws, what relative risk factor, from a legal point of view, should be assigned to each of the categories and their subdivisions?

RETROFIT VS. REPLACEMENT

Noise abatement of existing machinery may take the form of add-on modifications (retrofit) to the existing machine, or replacement of the machinery with new, quieter models. It would appear that a logical approach would be to first phase out older equipment as a noise abatement measure. While this approach may be correct from an economic or operations viewpoint, the opposite approach would have preferable acoustical results. In considering a long term noise abatement approach, it is the newer items of machinery which will create noise problems for many years, and which should be attacked with highest priority. The noise problems generated by equipment with an expected short remaining

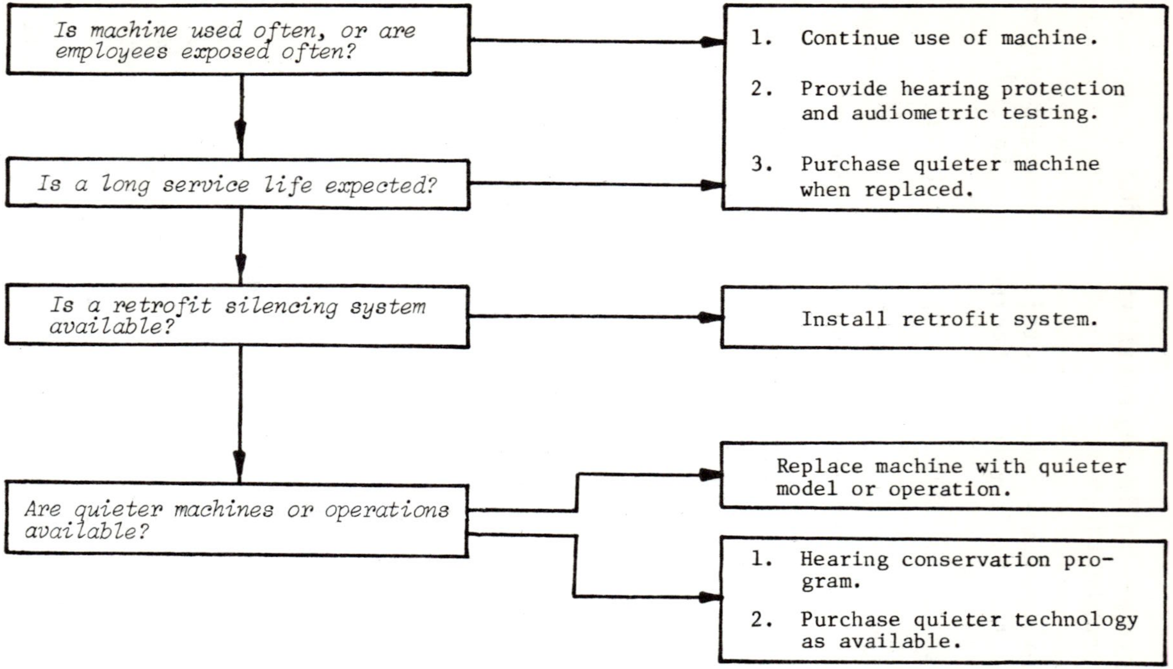

FIGURE 8.2. **Decision matrix for determining noise reduction approach for existing machines.**

service life will be abated automatically if noise specifications are enforced with new product purchases. A decision matrix for noise reduction approaches is presented in Figure 8.2.

FORMS FOR EVALUATING PRIORITIES

Establishing noise abatement priorities requires an understanding of the hearing risks as well as the economics involved. Figures 8.3 and 8.4 are forms which may be used in establishing priorities for noise projects.

NOISE ABATEMENT
PROJECT EVALUATION SUMMARY

CALCULATED:

Return on Investment ________________ %

Pay Back Period ________________ months

Other ________________ ________________

NOISE LEVEL:

Present ________ Projected After Project Implemented ________

PROBLEMS:

Product Quality ________________

Product Yield ________________

Production Rate ________________

Safety ________________

Pollution ________________

Maintenance: Manpower/Materials ________________

Utilities ________________

Working Conditions ________________

Employee Attitude ________________

Community ________________

Other Benefits/Problems Related to Implementation:

____

____

COMMENTS: ________________________________

PROJECT RATING: ________________________________

PLANNED AUTHORIZATION REQUEST DATE: ________________________________

FIGURE 8.3. Sample form for noise project evaluation.

NOISE ABATEMENT
CAPITAL PROJECTS SUMMARY

PROJECT NUMBER	PROJECT DESCRIPTION	EFFECTIVENESS (E)	CAPITAL COST (C)	E/C	RATE OF RETURN	PRIORITY	STATUS

FIGURE 8.4. Sample form for noise capital projects.

REFERENCES

1. Clark, E. O., "Determining Priorities of Current and Long-Range Pollution Control Projects," presented to the International Pollution Engineering Congress, October 22-25, 1973.

Chapter 9
EFFECTIVE NOISE MANAGEMENT

Three factors provide the primary impetus for control of industrial noise:

1. OSHA compliance.

2. Hearing loss and worker's compensation litigation.

3. Working environment and production efficiency.

It is the responsibility of management to work toward an effective application of these requirements in a manner which optimizes economic benefits to the company involved.

CRITERIA FOR NOISE ABATEMENT COMMITMENTS

The first step in defining a management commitment to noise control is the establishment of objectives. Fundamental objectives may include:

- Legal compliance
- Economic benefits
- Improved employee environment

Each proposed engineering or administrative control for noise abatement should be reviewed for consistency with the established objectives. A set of objectives such as the following should be established to review each proposed project prior to approval:

1. A noise control program should be selected only if the benefits to the firm are in excess of the estimated costs and if the lives and social security of the people are not otherwise adversely affected.[1]

2. A noise control program should be selected if conditions exist which potentially endanger the physical or mental health of employees, and if such a program would eliminate or decrease the severity of the condition.

3. Noise control programs should be selected if required by law.

MANPOWER RESOURCES

Six resource communities, each offering varied viewpoints and fields of expertise, may be utilized for any industrial noise abatement program:

1. In-house personnel
2. Professional consultants
3. Product manufacturers
4. Academic community
5. Professional organizations
6. Government agencies

Management planning must include selection of the optimum technical team from these resource areas to insure effective results. It is requisite in any noise control program that top level management have both a knowledge of noise control requirements and developments, and reliable personnel to whom they may turn for technical, medical, and legal solutions.

MANAGEMENT REQUIREMENTS

Even where responsibility is delegated, the manager should have an understanding of the following:

1. The full legal implications of noise exposure.

    a. OSHA requirements (see Chapter 2).

    b. State worker's compensation statutes (see Chapter 5).

2. The basic engineering approaches.

With a fundamental understanding in these two areas, the manager can rely on in-house or outside resources for detailed information, and can delegate development of design solutions to design professionals or in-house engineers.

PERSONNEL

The type of personnel structure to be established for implementing a noise abatement program is dependent upon a firm's necessity to employ permanent noise control engineering expertise. Three basic organizational structures are shown in Figure 9.1 and are discussed below.

HIRING AN IN-HOUSE SPECIALIST

When the noise control program is anticipated to require several years of concentrated effort, a fulltime permanent noise control specialist may be employed. To function as a self-reliant technical resource center, this individual would require probably two years of

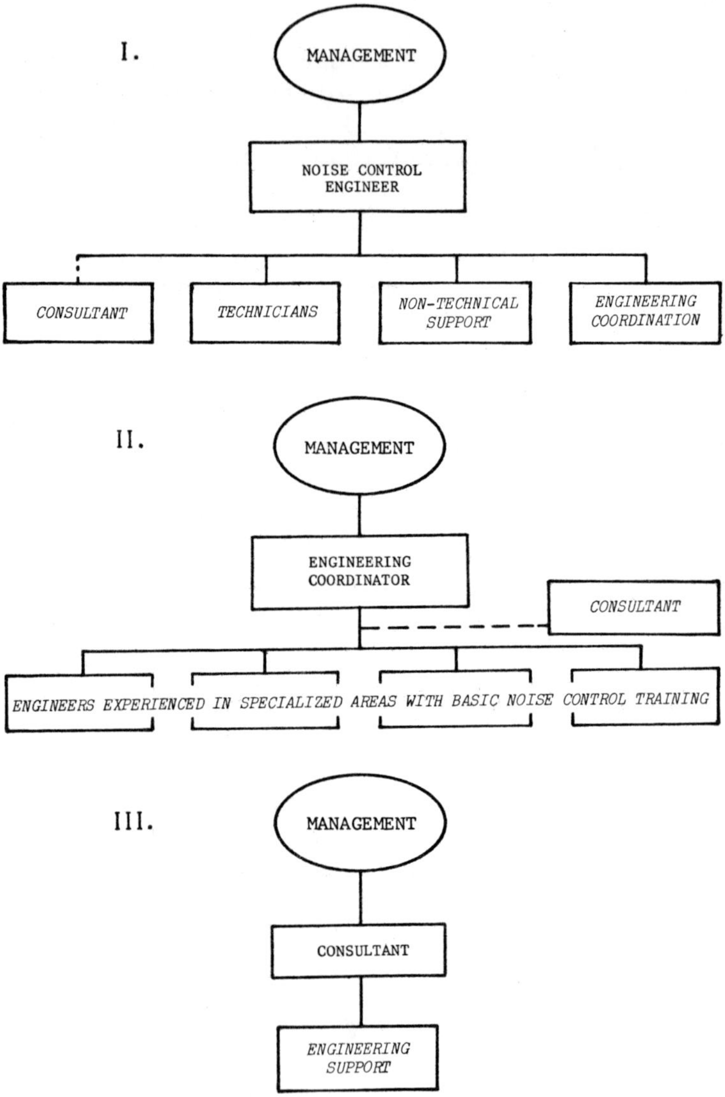

FIGURE 9.1. Responsibility delegation for a noise control engineering program.

noise control experience plus some type of acoustical educational background. It is generally not possible to develop an expert (in noise control or any other field) through training alone. Experience is necessary to develop problem solving abilities and knowledge of practical materials for implementation. If a noise control expert is hired, other plant engineers need to have only a "noise control awareness" rather than actual acoustical expertise. An outside consultant may be utilized for brainstorming or conceptual design in conjunction with the in-house expert to provide independent design review and to compliment in-house capabilities with specialized expertise (legal testimony, measurements requiring special instrumentation, etc.).

UTILIZING EXISTING IN-HOUSE PERSONNEL

Existing engineering personnel may be provided with a basic training in noise control to accomplish short range noise control programs. These engineers may gain knowledge through short training courses (several good one-week seminars are offered throughout the country; see Reference 2 for a partial listing), and through independent study. Several engineers may be trained simultaneously and can be assigned noise control projects relating to their area of specialization, working together in coordinating technical information. An outside consultant may be retained to conduct in-house training sessions or to work directly with engineers in various areas of noise reduction design.

CONTRACTING AN OUTSIDE CONSULTANT

Where it is anticipated that noise control objectives may be accomplished with a short term design effort, and where minimum reorganization is desired, an outside consultant may be utilized in lieu of development of in-house expertise. The consultant would require both management guidance to direct his efforts toward achievement of plant objectives and goals, and technical support to coordinate his designs with production requirements and to direct implementation of his recommendations. The decision to retain an outside acoustical firm, as opposed to utilizing in-house capabilities, would involve the same decision process as consideration of contracting outside legal or accounting services.

COST CONSIDERATIONS

In considering the relative costs of the three approaches, annual salary requirements for qualified noise control engineers generally range from \$20,000 to \$30,000. Training seminars normally cost from \$200 to \$500 per person plus expenses to the seminar location; however, the greatest cost would probably be the time required for engineers without acoustical experience to develop efficient noise control problem-solving expertise. Fees of consulting firms typically range from \$25 to \$60 per hour.

CORPORATE MANAGEMENT

An additional dimension of coordination is required for large corporations involving multiple plant operations. Optimum economic benefits will be achieved with proper utilization of corporate resources.

It is generally most convenient for the corporate office to assume responsibilities which involve areas of general applicability to all plants. Coordination of the noise control efforts of each plant by the corporate office is necessary to avoid duplication of effort.

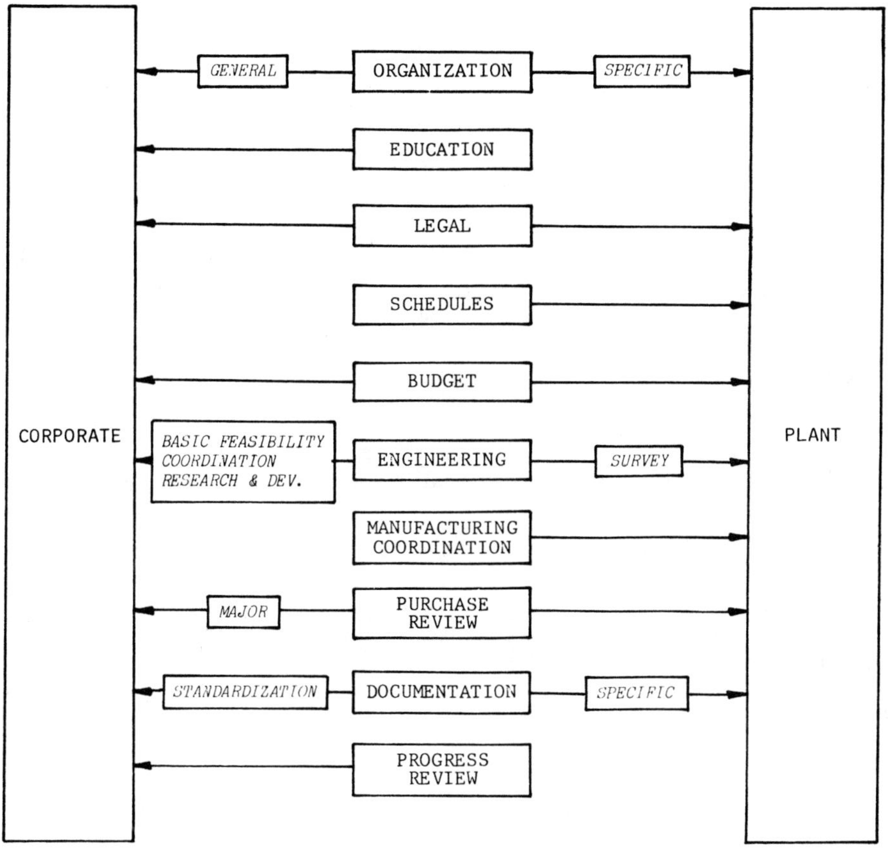

FIGURE 9.2. Corporate responsibility flow chart.

In cases where technical and managerial resourses exist at both the corporate and plant level, the responsibility assignments may be delegated as suggested in Figure 9.2.

NOISE CONTROL IN NEW PLANTS[3]

The complexity of the design of new industrial facilities requires the teamwork of several disciplines. Until recently, noise control was not a separate discipline in the engineering task force. Hence, each group would be responsible for noise problems related to its specific task. In many instances, this lack of noise coordination has led to an overdesign or underdesign with respect to noise. The noise control activity needs to be properly managed in order to insure that technical objectives, project costs and schedules are met. Unlike most other engineering activities, noise control transcends the traditional design boundaries. This unique aspect makes the management of noise control for industrial "grass roots" design a formidible task.

The management of noise control involves three key areas, namely: coordination of noise control activities, schedule awareness, and cost consciousness. Let's look at each of these areas.

COORDINATION OF NOISE CONTROL ACTIVITIES

The basic engineering organization structure for design of a new industrial facility which takes into account noise control is illustrated in Figure 9.3. For simplicity, construction, procurement, expediting, inspection, cost control, and senior management have not been included. Noise control is involved in some aspect of

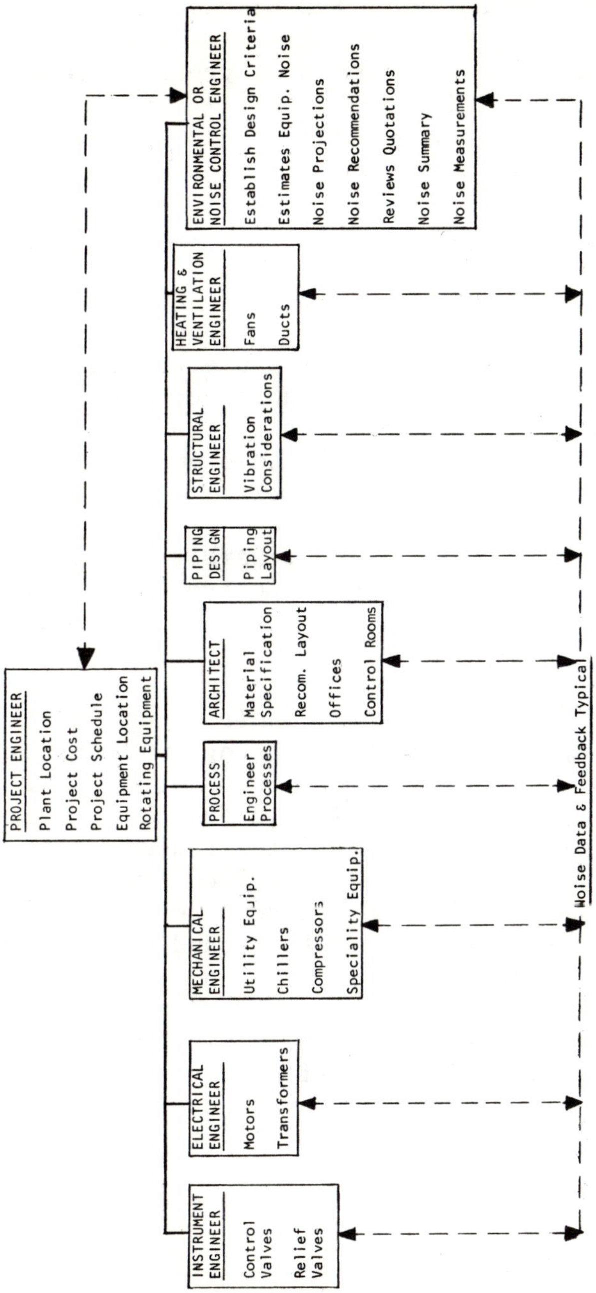

FIGURE 9.3. **Designation of noise control responsibility for new plant design.**

each discipline's work as indicated below:

1. *Architectural.* The architectural group is responsible for the specification of building materials and systems such as walls, floors and windows. These systems may serve to isolate noise sources from noise sensitive areas. In addition, the architectural design responsibilities include the auditorium, conference rooms, offices, plant location, and interior space location.

2. *Mechanical.* The mechanical or project group is responsible for the selection and location of process and utility equipment. Large items of mechanical equipment such as chillers and compressors are potential noise offenders.

3. *Instrument.* This group is responsible for instruments such as control valves. Control valves are potential noise producers and must be selected properly to insure minimum noise impact.

4. *Heating, Ventilation and Air Conditioning (HVAC).* The HVAC group specifies the heating, ventilation and air conditioning equipment required for the plant. The type of fans selected and their location will greatly affect the noise radiated to residences, process areas, and office areas. The type of diffusers selected and duct routing will affect background noise of offices and speech privacy between rooms.

5. *Electrical.* This group specifies electrical equipment such as power transformers, motors, and lighting. The low frequency hum and vibration of transformers may present annoyance problems for office areas or adjacent residences. Noise from motors may create problems in process or production areas.

6. *Structural.* The structural group is responsible for the structural steel and concrete specifications. Proper selection of these materials will minimize vibration caused by equipment excitation.

7. *Piping and Plant Layout.* This group determines routing and location of piping and mechancial equipment. Noise generating equipment must be routed away from noise-sensitive offices. For example, a noise-radiating pipe which passes above an office ceiling may create an annoyance. Outdoor equipment locations which take advantage of natural barriers may ease some noise problems.

8. *Process.* The process group affects industrial plant noise by the recommendation of process design. Sometimes a quieter process can be substituted for a noisier one.

9. *Project.* The project engineer is responsible for piping and instrument diagrams (P&ID), and overall coordination. The P&ID reflects all equipment purchased and should reflect vent and equipment silencers. He coordinates all aspects of the project, such as plant location. Plant location should not only be based on availability of resources, transportation, labor and material costs, but also on noise considerations.

10. *Noise Control.* The position of noise control engineer has evolved to coordinate noise data and to insure a consistent approach to noise control. The noise control activity may be a part of an overall environmental group. The interaction and coordination between the noise control engineer and the various disciplines is indicated by the dotted lines in Figure 9.3. The responsibilities of the noise control engineer include:

a. Analyzing equipment noise data.

b. Identifying potential noise offenders.

c. Implementing corrective action.

d. Reviewing equipment quotations *prior to purchase.*

e. Recommending plant and equipment location.

SCHEDULE AWARENESS

Noise control considerations affect virtually every aspect of design. Due to the effects of noise control on engineering, it important to account for noise control activities in the overall project schedule. Unless noise control considerations are made at the proper time, delays in noise engineering and procurement will occur. Figure 9.4 shows a typical schedule for noise control and the sequence of events of the various noise control activities. Noise control activities should commence at the beginning of the project to insure that key noise areas such as plant location, road access, equipment location, and construction materials are not overlooked. Another important consideration is estimation of equipment noise levels, so that it may be established which equipment should be silenced and to what degree. This information must be available prior to equipment purchase in order that the effect of each noise contributor may be analyzed.

COST CONSCIOUSNESS

There are three primary cost concerns facing the noise control engineer:

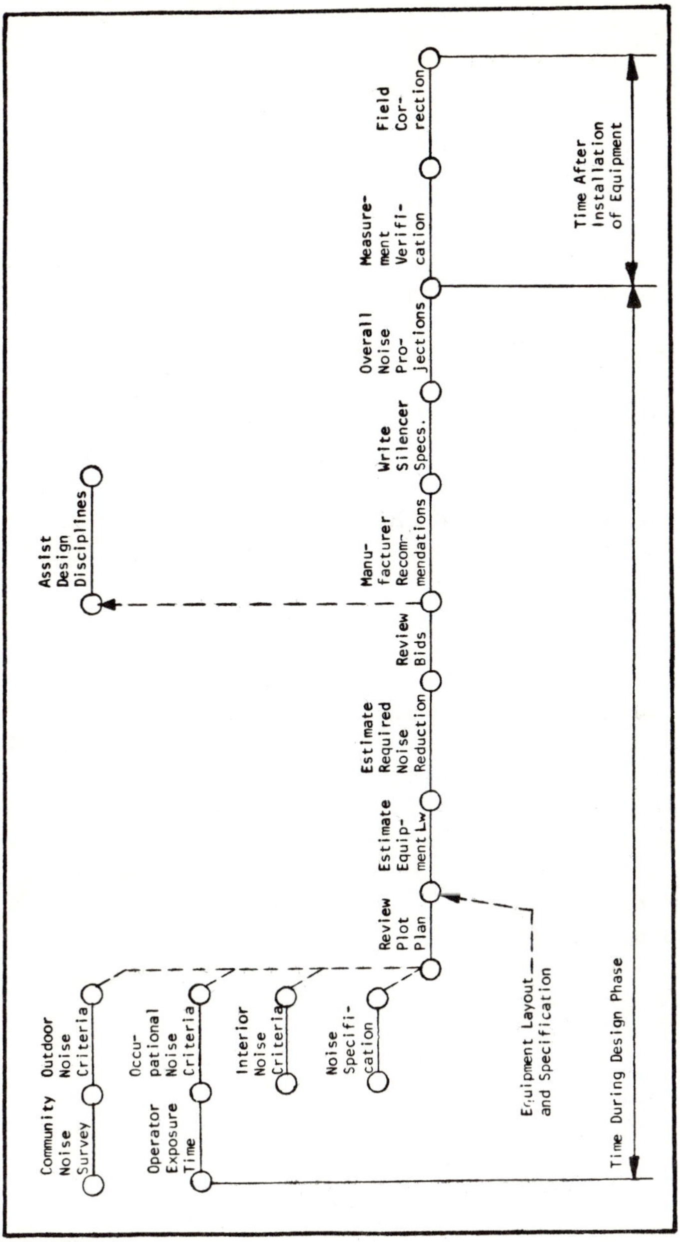

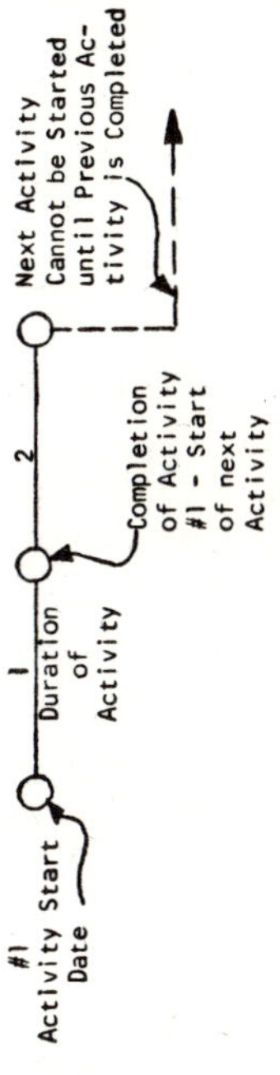

FIGURE 9.4. Schedule of noise control activities for plant design.

1. The effect of noise objectives on noise control costs.

2. The identification of noise control costs.

3. The incorporation of silencing provisions.

Let's look at each of these concerns.

Effect of Noise Objectives on Noise Control Costs

Criteria related to community and occupational noise exposure must be carefully selected in order to optimize results from noise control expenditures. The practice of arbitrarily setting maximum permissible noise levels must be avoided. For example, setting 85 dBA as a design goal rather than 90 dBA may more than double noise abatement costs. This cost must be carefully weighted against projected future trends in legislation, reduction of hearing loss risk, and other noise control benefits.

Community noise objectives must be carefully chosen based on existing state and local legislation, possible future legislation, and existing ambient sound levels as established by a noise level survey. The selection of an overly conservative environmental noise objective will result in unnecessary abatement costs; however, the selection of a liberal criteria may result in very costly future measures.

Identification of Noise Control Costs

Noise control costs are important to identify from a "target" budget and tax incentive point of view. External silencers and enclosure costs are easy to identify, but many noise costs are buried in other aspects of design. An illustration is the isolated control room. If it is required for reasons *other* than noise, it should *not* be included in the noise budget. If noise control is the primary reason for its existence, then it should be included.

Incorporation of Silencing Provisions

Incorporating silencing provisions early in the design stages of a new plant may reduce silencing costs significantly. For example, the proper orientation of a cooling tower during initial plant construction may eliminate the requirement for a noise barrier costing several thousand dollars if required after start-up. Similarly, "quiet line" motors may be specified which minimize noise through cooling fan blade design. As illustrated in Table 9.1, the specification of these motors may eliminate future requirements for an external housing which would cost several hundred dollars.

In other cases, it may not be evident if noise problems will occur, and there may be little cost penalty for later installation. Here, it may be wise to delay other noise control items until after the plant start-up. As an example, piping noise can generally only be predicted to within ±5 dBA. Since acoustical insulation may be installed easily after a plant is in operation, implementation of this noise control project may wait until after start-up.

Noise control for new plant construction must take into account that noise problems can be prevented by the engineer evaluating the *total plant design* from early conception phases through plant start-up.

HIRING A CONSULTANT

An acoustical consulting firm may be retained to supplement in-house capabilities in the development and

TABLE 9.1

MOTOR COST ANALYSIS*

1800 RPM TEFC Motor	Motor HP	Sound Power Level, dBA	Motor Costs	External Housing to Lower Noise Level 12 dBA	Savings
Standard	100	90	$2282	$1500	-
Quiet Line	100	78	$2314	-	$1368

* Based on 1974 motor costs including discount.

implementation of a noise abatement program. The role of the consultant may be to provide expertise not available in-house, or to minimize the impact of a noise control program on existing operations.

The advantages and disadvantages of hiring a consultant are presented in Table 9.2.

CONSULTANT'S FEE

The consulting fee of two firms for the same noise problem may differ by more than an *order of magnitude*. The primary difference may not be in the consultant's qualifications or profit motives, but in how the scope of work is defined. The scope of work may include the following items to any degree of depth:

1. Noise exposure survey.
2. Engineering feasibility study.
3. Conceptual design development.
4. Detailed design.
5. Acoustical management consultation.
6. In-house design review.
7. Technical support related to legal compliance aspects.
8. Turnkey service.

For example, one consulting firm veiwed a particular paper mill's noise problem as requiring an engineering feasibility study with a consulting fee of $4,000, while a second consulting firm proposed an in-depth

TABLE 9.2

NOISE CONSULTANT CHECKLIST

Advantages	Disadvantages
1. The consultant offers a noise expertise which the company usually does not possess. He may have experience with similar or identical problems. The consultant will generally have a broader base of information regarding such items as commercially available materials and feasibility of noise solutions.	1. The consultant may lack familiarity with plant operations. He may not give proper attention to non-acoustical requirements such as productivity, maintenance, fire codes, and FDA requirements.
2. The consultant helps to level out peak in-house manpower requirements. When the noise project is completed, no exces in-house manpower problem exists.	2. The consultant would not be on site if problems should develop subsequent to the design development.
3. The consultant's fee may be lower than in-house manpower costs when overhead, efficiency and employee benefits are considered.	3. The consultant's total fee may be higher than in-house manpower costs due to the depth of study and the consultant's hourly charge.

TABLE 9.2 (Continued)

Advantages	Disadvantages
4. The consultant offers an independent outside viewpoint which is desensitized to in-plant politics and manufacturers' products. OSHA may view the opinion of an outside consultant more openly than that of in-house personnel.	4. The consultant may recommend overdesigned and expensive solutions to insure the success of his efforts. He may not hold the company's interests in as high a regard as in-house employees. Conflicts may also arise with in-house personnel if the consultant is not adept at working with people.
5. The consultant may have an established rapport with governmental agencies involved in compliance programs.	5. The consultant may have a poor track record in his dealings with governmental agencies.

noise exposure study for a fee of $25,000. The feasibility study was undertaken, and satisfactory results were achieved. While the problem being attacked was identical for both consultants, the difference lay in how the consultants perceived the company's requirements, and also in the technical orientation of the firms.

Prior to hiring a consultant, a company should define what it perceives as its own requirements in order not to be "oversold" with an unnecessarily large scope of work. At the same time, the company should give careful consideration to the consultants' suggestions regarding improvement of the scope of work.

SELECTION AND MANAGEMENT OF AN OUTSIDE CONSULTANT

The following guidelines are offered to assist in selecting and managing a consultant:

1. The basic philosophy of the consultant should be in agreement with that of the client firm with regard to general solution approaches, goals, economics, legal compliance, ethics, etc.

2. Hourly fees are generally of only secondary importance; only the total fee and final results should be considered. Some consultants will require more time to accomplish certain projects. Thus, the total consulting fee from one firm may be less than that of another, even though the latter firm has lower hourly billing rates.

3. One experiences diminishing returns when retaining any expert. The consultant will generally establish fundamental approaches and accomplish the conceptual aspects of a project during the initial portion of the project; the remaining time will be

spent on more straightforward analysis and design.

4. The consultant should be provided with a written statement of a firm's objectives and goals. Consultants go wrong most often when they are misguided.

5. The results achieved by a consultant will be proportional to the assistance with which he is provided. He will be most motivated when he recognizes a sincere management commitment toward the noise control program, and his time will be best utilized when he is provided with all technical information necessary for his assessment or design.

6. Beware of consultants operating without sufficient background to provide competent problem-solving services. These firms often perform noise survey services rather than offering complete services.

7. One should consider consultants outside of the immediate geographic region, as well as local firms. Transportation expenses may constitute only 5-10% of a project fee, while the total fee may vary over 100%.

8. Past clients of each consulting firm being considered should be contacted with regard to satisfaction with the consulting services.

9. The experience of the consultant as an expert witness should be checked if the project involves a contested OSHA citation and trial is anticipated. Is he a winner, loser, or novice?

10. One should not be overimpressed with a consultant's academic credentials unless supplemented with actual experience. Academic training may be

directed toward theoretical research. The consultant must also have applied, problem-solving abilities.

11. Several consulting firms should be considered for a project. Qualifications and fees may vary significantly among firms.

12. Manufacturing firms offering "free" consulting services should be considered cautiously because:

    a. Their experience may be limited.

    b. The services are generally not "free" but are covered by an increased cost in materials.

    c. They generally only recommend their own products, which may not provide the best or most economical solution for an application.

13. When dealing with larger firms, the qualifications should be sought for the individual consultants who would be involved in the project. A qualified firm may utilize inexperienced consultants for actual project involvement.

14. The use of larger and recognized consulting firms should not be overlooked because of possible high fees. Many top-rated firms cost no more than less experienced firms.

15. When consulting projects are undertaken on a time plus expenses basis, a "not to exceed" fee should be established.

16. Many qualified consultants are too busy or understaffed to provide service to meet a client's needs. A good indication of a firm's promptness is the time it takes for response to your inquiry.

17. A consultant should be viewed for his sensitivity toward economic and production requirements as well as acoustical expertise. He should have a thorough knowledge of commercially available systems and materials.

18. Insurance firms often provide noise surveys for worker's compensation policy holders; however, they do not generally provide engineering design services. Their services are intended to identify hearing damage risk, and should not be confused with acoustical design consulting services.

19. It is not reasonable to expect a guarantee from a consultant that a specified noise level would be reached based on his recommendations. In any scientific study there are many variables which may result in a degree of uncertainty in the response of any system. However, the consultant should be responsible for design mistakes.

20. The consultant may be considered to develop in-house noise control capabilities as well as to develop noise abatement designs. In this role he may present in-plant seminars or train engineers on an informal basis.

21. The consultant should communicate to OSHA, plant personnel, and all others only through management. The role of the consultant is as an advisor to his client, and not to establish policies or make commitments for the firm.

22. Any documents submitted to OSHA relating to compliance plans prepared by the consultant should be identified as such.

23. All contracts or retainer agreements with the consultant should be severable if the client is not

satisfied, provided adequate notice is given.

STRATEGIES, SCHEDULING & RELATED ECONOMICS

Solutions to noise problems may be categorized by difficulty and cost into four classes:

Class	*Definition*
I	Solutions evident with little or no cost
II	Solutions evident and economical
III	Solutions evident and costly
IV	Solutions not evident

A company can approach these problems with various levels of aggressiveness or various strategies. The cost and effectiveness will vary with each approach. Let's take a look at each of the problem categories and related strategies.

CLASS I

A policy of "noise awareness" may be established with little or no cost. This program would include two elements:

1. Engineers would be educated or trained in noise reduction techniques for new designs and operations.

2. Noise control specifications would be included in new product purchasing procedures.

This policy would result in a gradual decrease in the

plant noise exposure, since plant operations and equipment undergo a continuous evolutionary process involving production improvements and machinery replacement. Associated wtih these changes will be reduced noise levels due to lower noise levels of new equipment as manufacturers respond to market demands and in-house "noise awareness." Results will be gradual and costs will be low.

CLASS II

A policy may be adopted to implement noise abatement measures with reasonable costs within a reasonable time schedule. Solutions with excessive costs may be further studied. As the state-of-the-art advances, the cost of these solutions may be reduced, and implementation may be possible. This kind of program would have moderate costs and effective results.

CLASS III

A commitment may be made to implement all available noise control measures without regard for costs, and to initiate research and development in an effort to find solutions for all other noise problems. Noise reduction will be accomplished at the highest possible level, and associated costs will also be maximized.

CLASS IV

Where noise control solutions are not technically feasible, economic alternatives do not exist. While direct noise abatement costs are saved, the company ultimately faces the environmental and economic consequences of their noise problem unless alternate technical approaches are developed in the future.

CHOOSING THE BEST SOLUTION

Guinter[4] has outlined the basic management technique for problem solving as related to noise reduction design.

> [First the manager must] define performance goals, such as the least change acceptable, the desired change, and factors defining feasibility. [Next,] proceed to generate without criticism, by brainstorming or other means, a large list of potential solutions. Alex Osborne[5] says that the first 14 to 17 ideas tend to be stereotypes; you should cultivate many more than that....[and] judge them as possible, unlikely, or presently impossible. If you discover what "went wrong" with the malfunctioning but previously satisfactory situation, you may be satisfied simply to restore the system to its former condition.
>
> If you seek a new solution, rank the potential solutions. Kepner and Tregoe[6] suggest a worksheet and a checklist which forces one to define the "must" requirements, and to rank the "want" features against which potential solutions must be scored. Resources required, and the risk of undesirable consequences are also rated. The technique discloses significant differences among the ideas if such exist, and makes the basis for judgement visible for discussion.
>
> Next, break the most desirable solution and one or two alternatives into readily accomplished steps, defining what must be done in each, means for rating its accomplishment, and estimate the time and resources needed. Study whether the expected result will justify the

estimated cost. Encourage the participants in the program to adopt realizable goals.

THE COMPLETE NOISE ABATEMENT PROGRAM

To achieve a successful noise abatement program, the manager must insure that proper emphasis is directed toward *all* elements associated with the program. Ignoring any single management responsibility may result in serious undesirable consequences. For example, a plant which successfully and economically reduces noise levels from 110 dBA to 80 dBA may not have an acceptable noise program. Why? They may be liable for thousands of dollars in worker's compensation claims if the audiometric testing phase of their program has been overlooked. It is all to easy to concentrate on the major items of noise abatement and ignore equally important items such as documentation and economic analysis.

No manager wants his plant to become a statistic. He does not want his firm to be one of the .05% of American businesses cited each year by OSHA for a noise violation, nor does he desire his noise budget to be a significant portion of the $10.5 billion required of American industry to comply with OSHA's 90 dBA standard. At the same time, he must strive to avoid allowing excessive noise levels to jeopardize the health, safety and morale of his employees. It has been the objective of this handbook to provide the manager with the background necessary to organize his efforts to meet these challenges. As a final measure, the following outline itemizes the primary elements of an industrial noise abatement program and may be used as a checklist to insure that all elements of a complete noise abatement program have been considered.

A. ASSESSMENT OF NOISE EXPOSURE

1. Survey dBA levels and exposure times for OSHA compliance determination.

2. Measure noise spectrum and vibration levels, and obtain other information required for noise control engineering.

B. ACHIEVEMENT OF OSHA COMPLIANCE

1. Implement a hearing conservation program.

   a. Hearing protective devices
   b. Audiometric testing

2. Establish whether noise control engineering is feasible.

   a. Technical considerations
   b. Production considerations
   c. Economic considerations

3. Develop a schedule for noise abatement engineering.

4. Document all items.

C. NOISE CONTROL ENGINEERING

1. If engineering controls appear to be feasible, assess all potential approaches to noise abatement.

   a. Acoustical enclosures and barriers
   b. Sound absorptive treatment
   c. Vibration isolation

d. Vibration damping
e. Mufflers
f. Machine redesign
g. Process substitution
h. Administrative controls

2. Implement the optimum engineering solution.

D. ECONOMIC ANALYSIS

1. Assess potential economic benefits.

a. Reduced risk of worker's compensation payments
b. Improved work environment
c. Improved morale, reduced absenteeism and turnover
d. Increased production
e. Energy savings
f. Governmental incentives

2. Assess projected costs.

a. Basic materials and installation costs
b. Operational and production costs

E. ORGANIZATION

1. Establish a noise abatement policy.

2. Establish priorities.

3. Coordinate manpower.

a. Corporate involvement
b. In-plant engineers
c. Acoustical consultants

d. Audiometric testing firms or personnel
e. Legal council
f. Material suppliers and installers

4. Establish scheduling.

5. Provide for documentation of progress.

REFERENCES

1. The Rivers and Harbor Act of 1902.

2. Baker, W. L., "Survey of Short Courses in Acoustics," *Journal of the Acoustical Society of America*, Vol. 56, No. 3, September 1974, pp. 1025-1028.

3. Thumann, A., "The Management of Noise Control for New Plants," presented at the Fifth Annual Environmental Engineering and Science Conference, Louisville, Kentucky, March 3-4, 1975.

4. Guinter, J. M., "Planning Development to Reduce Product Noise," Proceedings of Noise-Con 73, pp. 370-375.

5. Osborn, Alex F., *Applied Imagination*, Charles Schribners Sons, New York, 1953, 1957.

6. Kepner, C. H., and Tregoe, B. B., *The Rational Manager*, McGraw-Hill, New York, 1965.

7. Guinter, J. M., "Action Plan for Product Noise Reduction," *Sound and Vibration*, Vol. 6, No. 12, December 1972, p. 18.

Appendix A

Appendix A

OSHA NOISE STANDARD

§ 1910.95 Occupational noise exposure.

(a) Protection against the effects of noise exposure shall be provided when the sound levels exceed those shown in Table G–16 when measured on the A scale of a standard sound level meter at slow response. When noise levels are determined by octave band analysis, the equivalent A-weighted sound level may be determined as follows:

Figure G-9

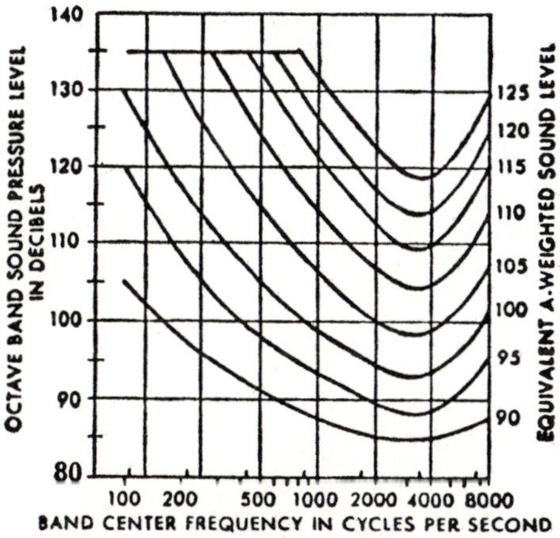

Equivalent sound level contours. Octave band sound pressure levels may be converted to the equivalent A-weighted sound level by plotting them on this graph and noting the A-weighted sound level corresponding to the point of highest penetration into the sound level contours. This equivalent A-weighted sound level, which may differ from the actual A-weighted sound level of the noise, is used to determine exposure limits from Table G–16.

[1910.95 amended at 39 FR 19468, June 3, 1974]

(b) (1) When employees are subjected to sound exceeding those listed in Table G–16, feasible administrative or engineering controls shall be utilized. If such controls fail to reduce sound levels within the levels of Table G–16, personal protective equipment shall be provided and used to reduce sound levels within the levels of the table.

(2) If the variations in noise level involve maxima at intervals of 1 second or less, it is to be considered continuous.

(3) In all cases where the sound levels exceed the values shown herein, a continuing, effective hearing conservation program shall be administered.

TABLE G–16—PERMISSIBLE NOISE EXPOSURES [1]

Duration per day, hours	*Sound level dBA slow response*
8	90
6	92
4	95
3	97
2	100
1½	102
1	105
½	110
¼ or less	115

[1] When the daily noise exposure is composed of two or more periods of noise exposure of different levels, their combined effect should be considered, rather than the individual effect of each. If the sum of the following fractions: $C_1/T_1+C_2/T_2+\dots C_n/T_n$ exceeds unity, then, the mixed exposure should be considered to exceed the limit value. *Cn.* indicates the total time of exposure at a specified noise level, and *Tn* indicates the total time of exposure permitted at that level.

[1910.95 Table G16 amended at 39 FR 19468, June 3, 1974]

Exposure to impulsive or impact noise should not exceed 140 dB peak sound pressure level.

Appendix B

Appendix B

REVISED OSHA NOISE STANDARD

DEPARTMENT OF LABOR

Occupational Safety and Health Administration

[29 CFR Part 1910]

[Docket No. OSH-11]

OCCUPATIONAL NOISE EXPOSURE

Proposed Requirements and Procedures

I. *Background.* On August 14, 1972, the National Institute for Occupational Safety and Health (NIOSH), Department of Health, Education, and Welfare, provided to the Department of Labor a criteria package, "Occupational Exposure to Noise" (HSM 73-11001), in accordance with section 20(a)(3) of the Occupational Safety and Health Act of 1970 (84 Stat. 1610, 29 U.S.C. 669). Thereupon, the Assistant Secretary of Labor for Occupational Safety and Health appointed a Standards Advisory Committee on Noise under section 7(b) of the Act (84 Stat. 1597, 29 U.S.C. 656). The purpose of this Committee was to obtain and evaluate additional recommendations from labor, management, government, and independent experts. The Committee in its deliberations considered 138 written comments directed to it by interested parties, as well as numerous oral presentations. It transmitted recommendations for a revised standard to the Occupational Safety and Health Administration (OSHA) on December 20, 1973.

OSHA has evaluated the information and recommendations contained in the NIOSH criteria package and the Advisory Committee's report. The OSHA proposal set forth below incorporates these recommendations with several additions and modifications.

II. *Exposure limits.* The current standard on occupational noise exposure limits an employee's exposure to 90 dBA, as an eight hour time weighted average. NIOSH indicated a need for reducing this eight hour exposure level to 85 dBA, but was unable to recommend a specific time period after which the 85 dBA noise level should become effective for all industry due to the unavailability of sufficient data relating to technological feasibility of this level. Therefore, NIOSH reluctantly concurs with the generally acceptable 90 dBA occupational exposure level for an eight hour day. The Advisory Committee's final report recommended retaining the 90 dBA permissible limit for an eight hour day.

The Environmental Protection Agency (EPA), basing its recommendations on a review of hearing impairment risk, has recommended that OSHA reduce the limit at least to 85 dBA. EPA, which reviewed the draft standard proposal under the authority of section 4(c)(2) of the Noise Control Act of 1972, further recommended that additional studies be undertaken to explore reducing the permissible level still further at some future date.

With regard to the risk of hearing loss, OSHA recognizes that comparatively more workers will be at lower risk at 85 dBA than at 90 dBA. However, we also recognize the technical feasibility problems and the economic impact associated with an 85 dBA requirement as reflected in the Bolt Baranek and Newman study and in the draft Environmental Impact Statement. Therefore, OSHA proposes to keep the level at 90 dBA until further empirical data and information on the health risk, feasibility, and economic impact indicate the practicality and necessity of an 85 dBA requirement. We feel that the present level, when coupled with a stringent hearing conservation program beginning at 85 dBA, will reduce the risk to an acceptable level during the period of further consideration.

OSHA is aware of several studies currently under way that may provide additional information for this determination. In addition, the audiometric testing program required by this proposal should provide even more information on this issue. OSHA also requests any available information that can help clairfy the question of permissible level.

Another consideration in determining the permissible exposure level is the appropriate doubling rate. This rate is the amount by which exposure intensity may be increased when exposure time is decreased. For example, the 5 dB doubling rate incorporated in the present standard and continued in the proposal would allow an exposure of 85 dBA for 16 hours; 90 dBA for 8 hours; and 95 dBA for 4 hours.

EPA recommended a doubling rate of 3 dB. While the 3 dB doubling rate is hypothetically correct for uninterrupted noise exposure, noise exposure in industry is normally interrupted since there are several breaks in the day's work. OSHA agrees with the Advisory Committee that the doubling rate should be adjusted to take into account the various breaks which occur in a workday. Therefore, OSHA believes that a doubling rate of 5 dB is more appropriate than 3 dB.

The present OSHA standard recommends that impact or impulse sounds not exceed a peak sound pressure level of 140 dB. The Advisory Committee suggested that this limit be made mandatory. OSHA has made an addition to the Advisory Committee's recommendation with respect to impulse noise exposure, because the actual exposure is a summation of the peak sound levels of the impulses and the number of impulses. OSHA proposes to limit exposure to impulses at 140 dB to 100 per day and to permit a tenfold increase in the number of impulses for each 10 dB decrease in the peak pressure of the impulse. For example, the number of impulses allowed at 130 dB would be 1,000 per day and the number of impulses allowed at 120 dB would be 10,000 per day. This proposal is in accordance with the criterion proposed by McRobert and Ward (H. McRobert and W. D. Ward, "Damage Risk Criteria: The Trading Relation Between Intensity and the Number of Non-Reverberant Impulses," "J. Acoustic. Soc. Am." 53: 1297–1300 (1973)).

III. *Controls*. The current standard states that feasible engineering and administrative controls shall be used to reduce noise exposure to within permissible levels. If such controls are not feasible or cannot reduce the sound levels to within permissible limits, then personal protective equipment shall be used to achieve compliance with the standard. This proposal continues the requirement that engineering and administrative controls be applied first to reduce noise to within permissible levels. The proposal also makes it clear that if engineering and administrative controls are not sufficient to reduce noise exposure to within permissible levels, such controls must nevertheless be used to reduce exposure to the greatest extent feasible and must be supplemented by personal protective equipment to achieve compliance. The proposal requires that all engineering and administrative controls be implemented, except where they are infeasible.

It is the opinion of OSHA that general use of hearing protection devices as a primary means of controlling noise exposure is not good industrial hygiene practice. It is not a satisfactory method of reducing noise exposure due to administrative difficulties commonly associated with the use of hearing protective devices. Thus, while hearing protective devices might technically afford the same protection as engineering controls, experience has shown that the protection afforded is diminished by the difficulty of management in requiring their use, workers' resistance to using them, and improper use and improper maintenance of such equipment. Accordingly, except in certain limited circumstances this proposal relies primarily on engineering and administrative controls to reduce employee noise exposure. OSHA has, however, received numerous recommendations to permit the use of hearing protective devices in lieu of expensive engineering controls to reduce the workplace noise level. Some claim that hearing protective devices are as effective as engineering controls in reducing noise exposure and may even have the added benefit of reducing worker exposure far below 85 dBA. Therefore, it has been suggested that the employer be given the choice of which method is utilized to achieve compliance with the standard. In view of the controversy surrounding the desirability of using protective devices to reduce employee noise exposure, OSHA will welcome and consider any submissions concerning the effective use of these devices.

IV. *Hearing conservation program*. Due to the wide variation in susceptibility to hearing loss from noise, it is not possible to determine a sound level exposure which will prevent all hearing loss in all members of an exposed population. The proposed requirement for a hearing conservation program that includes audiometric testing will prevent, or at least minimize, permanent noise induced hearing loss by identifying those workers especially susceptible to noise. Audiometric testing will be initiated for all employees with eight hour noise exposures of 85 dBA or higher. The audiometric testing program will also be required for those employees who wear personal protective equipment to reduce their noise exposures. Testing is necessary for these employees to assure that their hearing protectors are being used effectively.

OSHA believes that the audiometric testing program will detect any changes in hearing level in these workers so that the employer can adopt corrective action and inform employees before the changes become significant. In order for the results of such a testing program to be valid and meaningful, the audiometric environment and technique must be well standardized and stable over a sufficient number of years to represent a significant fraction of the employee's working life. It is also essential that these factors be reasonably identical from one employment to another. For these reasons, mandatory requirements are proposed for audiometric test rooms and the calibration of audiometers in the Appendix of this proposal.

V. *Monitoring*. Although the current standard requires exposures to be controlled within specified limits, it does not explicitly require monitoring of the sound level of the employee's surroundings nor measurement of the individual employee's resulting exposure. This proposal makes monitoring and measuring requirements explicit. It requires the employer to determine if any employee is exposed to an eight hour time weighted average of 85 dBA or above. If any employees are so exposed, the employer must identify such employees and measure their exposure. The proposal also prescribes the minimum acceptable accuracy for monitoring instruments.

VI. *Records*. The proposal requires the maintenance of records of the results of required measuring, monitoring and the calibration of the instruments used therein. Employers are required to retain employee audiograms for the duration of employment plus five years; retention of other records is required for five years. The proposed period of retention of these records reflects OSHA's evaluation of their future usefulness to employers, employees and the government.

In addition, the proposal implements the requirements of section 8(c)(3) of the Act (84 Stat. 1599, 29 U.S.C. 657), which concerns, among other things, employee access to monitoring records. Accordingly, the proposal contains provisions concerning the employee right to observe monitoring and provisions assuring employees and former employees access to records of noise exposure. In addition, the proposal would require that prompt written notification be given to any employee who has been exposed to noise in excess of the permissible limits. This notification must be accompanied by a statement of the corrective action being taken.

Accordingly, pursuant to section 6(b) and 8(c)(3) of the Williams-Steiger Occupational Safety and Health Act of 1970 (84 Stat. 1593, 1599; 29 U.S.C. 655, 657), Secretary of Labor's Order No. 12–71 (36 FR 8754), and 29 CFR Part 1911, it is hereby proposed to revise § 1910.95 of Title 29 of the Code of Federal Regulations as set forth below.

Written data, views, and arguments concerning the proposal may be mailed to Docket Officer, Docket OSH–11, Occupational Safety and Health Administration, Room 230, 1726 M Street, NW, Washington, DC 20210, on or before December 9, 1974. The data, views, and arguments will be available for public inspection and copying at the above address.

Pursuant to 29 CFR 1911.11 (b) and (c), interested persons may, in addition to filing written matter as provided above, file objections to the proposal and request an informal hearing with respect thereto in accordance with the following conditions:

(1) The objections must include the name and address of the objector;

(2) The objections must be postmarked on or before December 9, 1974.

(3) The objections must specify with particularity the provision of the proposed rule to which objection is taken, and must state the grounds therefor;

(4) Each objection must be separately stated and numbered; and

(5) The objections must be accompanied by a summary of the evidence proposed to be adduced at the requested hearing.

As proposed, § 1910.95 would read as follows:

§ 1910.95 Occupational noise exposure.

(a) *Application and purpose.* This section applies to occupational noise exposures in employments covered in this part. The purpose of this standard is to establish requirements and procedures that will minimize the risk of permanent hearing impairment from exposure to hazardous levels of noise in workplaces.

(b) *Definitions.* "Administrative controls" means any procedure which limits daily noise exposure by control of the work schedule. Hearing protectors do not constitute administrative controls.

"Assistant Secretary" the Assistant Secretary of Labor for Occupational Safety and Health, U.S. Department of Labor, or his designee.

"Audiogram" a graph or table of hearing level as a function of frequency that is obtained from an audiometric examination.

"Baseline audiogram" the first audiogram taken during employment with the current employer.

"Certified audiometric technician" an individual who meets the training requirements specified by the Intersociety Committee on Audiometric Technician Training (American Industrial Hygiene Association Journal, 27:303–304 (May–June 1966)) or who is certified by the Council of Accreditation in Occupational Hearing Conservation.

"Daily noise dose" (D) the cumulative noise exposure of an employee during a working day.

"dBA" (decibels—A-weighted)—a unit of measurement of sound level corrected to the A—weighted scale, as defined in ANSI S1.4–1971, using a reference level of 20 micropascals (2×10^{-5} Newtons per square meter).

"Director" the Director, National Institute for Occupational Safety and Health, U.S. Department of Health, Education, and Welfare, or his designee.

"Engineering control" any design procedure that reduces the sound level.

"Hearing level" the amount, in decibels, by which the threshold of audibility for an ear differs from the standard audiometric reference level.

"Peak sound pressure level" the peak instantaneous pressure expressed in decibels, using a reference level of 20 micropascals.

"Workplace sound level" the sound level measured at the employee's point of exposure.

"Impulse or impact noise"—a sound with a rise time of not more than 35 milliseconds to peak intensity and a duration of not more than 500 milliseconds to the time when the level is 20 dB below the peak. If the impulses recur at intervals of less than one-half second, they shall be considered as continuous sound.

"Significant threshold shift" an average shift of more than 10 dB at frequencies of 2000, 3000, and 4000 Hz relative to the baseline audiogram in either ear.

(c) *Permissible exposure limits.*—(1) *Steady state noise—single level.* (i) The permissible exposure to continuous noise shall not exceed an eight hour time-weighted average of 90 dBA with a doubling rate of 5 dBA. For discrete permissible time and exposure limits, refer to Table G–16a, which is computed from the formula in paragraph (c)(1)(ii) of this section.

TABLE G–16a

Sound level (dBA)	*Time permitted (hours-minutes)*	*Sound level (dBA)*	*Time permitted (hours-minutes)*
85	16–0	101	1–44
86	13–56	102	1–31
87	12–8	103	1–19
88	10–34	104	1–9
89	9–11	105	1–0
90	8–0	106	0–52
91	6–58	107	0–46
92	6–4	108	0–40
93	5–17	109	0–34
94	4–36	110	0–30
95	4–0	111	0–26
96	3–29	112	0–23
97	3–2	113	0–20
98	2–50	114	0–17
99	2–15	115	0–15
100	2–0		

(ii) Where Table G–16a does not reflect actual exposure times and levels, the permissible exposure to continuous noise at a single level shall not exceed a time amount "T" (in hours) computed by the formula:

$$T=\frac{16}{2^{[.2(L-85)]}}$$

where "L" is the workplace sound level measured in dBA on the slow scale of a standard sound level meter. The relationship between time and sound level is depicted in Figure G–9.

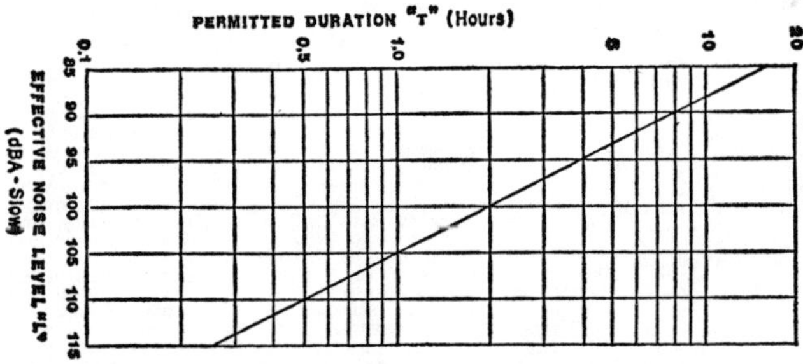

Figure G-9

(2) *Steady state noise—two or more levels.* Exposures to continuous noise at two or more levels may not exceed a daily noise dose "D" of unity (1) where "D" is computed by the formula:

$$D=\frac{C_1}{T_1}+\frac{C_2}{T_2}+\dots+\frac{C_n}{T_n}$$

where C is the actual duration of exposure (in hours) at a given steady-state noise level; and T is the noise exposure limit (in hours) for the level present during the time C, computed by the formula in paragraph (c)(1)(ii) of this section.

(3) *Maximum steady state noise level.* Exposures to continuous noise shall not exceed 115 dBA, regardless of any value computed in paragraphs (c)(1) or (c)(2) of this section.

(4) *Impulse or impact noise.* (i) Exposures to impulse or impact noise shall not exceed a peak sound pressure level of 140 dB.

(ii) Exposures to impulses of 140 dB shall not exceed 100 such impulses per day. For each decrease of 10 dB in the peak sound pressure level of the impulse, the number of impulses to which employees are exposed may be increased by a factor of 10.

(d) *Monitoring*—(1) *Duty*. Each employer shall determine if any employee is exposed to a daily noise level dose of 0.5 or above, and shall determine if any employee is exposed to impulse or impact noise in excess of the exposure permitted by paragraph (c)(4) of this section. Such determinations shall be made:

(i) At least annually, and

(ii) Within 30 days of any change or modification of equipment or process, or other workplace or work practice modifications affecting the noise level.

(2) *Procedure*. If determinations made pursuant to paragraph (d)(1) of this section reveal any employee exposure to a daily noise dose of 0.5 or above, or exposure to impulse or impact noise in excess of that permitted by paragraph (c)(4) of this section, the employer shall:

(i) Identify all employees who may be so exposed;

(ii) Measure the exposure of the employees so identified; and

(iii) Make all noise level measurements with the microphone of the sound measuring instrument at a position which most closely approximates the noise levels at the head position of the employee during normal operations.

(3) *Equipment*. (i) Measurements of steady state noise exposures shall be made with a sound level meter conforming as a minimum to the requirements of ANSI S1.4–1971, Type 2, and set to an A-weighted slow response or with an audiodosimeter of equivalent accuracy and precision. The unit of measurement shall be decibels re 20 micropascals A-weighted.

(ii) Measurements of impulse or impact noise exposures shall be made with a sound level meter conforming as a minimum to the requirements of the ANSI S1.4–1971, Type 1 or Type 2, with a peak hold capability or accessory. For peak hold measurements, the rise time of the strumentation) shall be not more than 50 microseconds. The decay rate for the peak hold feature shall be less than 0.05 decibels per second. The unit of measurement shall be decibels peak sound pressure level re 20 micropascals.

(4) *Calibration of equipment*. An acoustical calibrator accurate to within plus or minus one decibel shall be used to verify the before and after calibration of the sound measuring instrument on each day noise measurements are taken.

(5) *Observation of monitoring*—(i) *Duty*. The employer shall give employees or their representatives an opportunity to observe any monitoring of the noise levels in the workplace which is conducted pursuant to this section.

(ii) *Notification of employee right*. Written notice of the opportunity to observe the monitoring required by this section shall be prominently posted in a place regularly visited by affected employees and where notices to employees are usually posted. The employer shall take steps to insure that this notice is not altered, defaced, or covered by other material.

(A) The notice shall be posted at least three working days before monitoring is scheduled to occur.

(B) The notice shall list the time and place where monitoring will take place.

(C) The employer may require the employee or the employee representative to give advance written notification of intent to observe such monitoring.

(iii) *Exercise of opportunity to observe monitoring*. (A) When observation of the monitoring of the workplace for noise levels requires entry into an area where the use of personal protective devices is required, the employer shall provide and the observer shall use such equipment and comply with all other applicable safety procedures.

(B) Observers shall be given an explanation of the procedure to be followed in measuring the workplace noise level.

(C) Observers shall be permitted, without interference with persons performing the monitoring, to:

(*1*) Visually observe all steps related to the collecting and evaluating of the noise level data that are being performed at the time;

(*2*) Record the results obtained; and

(*3*) Have a demonstration of the calibration function tests of the monitoring equipment when the calibrations are performed at the worksite before monitoring; where the calibrations are not performed at the worksite, the techniques shall be explained.

(e) *Methods of compliance*. (1) Whenever employees are exposed to workplace sound levels exceeding those permitted by paragraph (c) of this section, engineering and administrative controls shall be utilized to reduce employee noise exposure to within permissible limits, except to the extent that such controls are not feasible. If such controls fail to reduce sound levels to within the permissible limits of paragraph (c) of this section, they shall be used to reduce the sound levels to the lowest level feasible and shall be supplemented by personal protective equipment in accordance with paragraph (f) of this section to further reduce the noise exposure to within permissible limits. Where the engineering and administrative controls which have been implemented do not reduce the sound levels to within the permissible limits of paragraph (c) of this section, the employer shall continue to develop and implement engineering and administrative controls as they become feasible.

(2) A program shall be established and implemented to reduce exposures to within the permissible exposure limit, or to the greatest extent feasible, solely by means of engineering controls. Written plans for such a program shall be developed and furnished upon request to authorized representatives of the Assistant Secretary and the Director.

(3) Exception. Hearing protectors may be provided to, and used by an employee to limit noise exposures in lieu of feasible engineering and administrative controls if the employee's exposure occurs on no more than one day per week.

(f) *Hearing protectors*. (1) Hearing protectors shall be provided to, and used by: (i) Employees receiving a daily noise dose between 0.5 and 1.0 (a daily noise dose of 0.5 is equivalent to an eight hour time weighted exposure of 85 dBA) if their audiograms show any significant threshold shift;

(ii) Employees who receive noise exposures in excess of the limits prescribed in paragraph (c) of this section: (A) During the period required for the implementation of feasible engineering and administrative controls; (B) in instances where engineering and administrative controls are feasible only to a limited extent; or (C) in instances where engineering and administrative controls have been shown to be infeasible.

(2) Hearing protectors shall reduce employee noise exposure to within the limits prescribed in paragraph (c) of this section.

(3) Procedures shall be established and implemented to assure proper issuance, maintenance, and training in the use of hearing protectors.

(g) *Hearing conservation*—(1) *General*. (i) A hearing conservation program shall be established and maintained for employees who:

(A) Receive a daily noise dose equal to or exceeding 0.5; or

(B) Are required to wear hearing protectors pursuant to paragraph (f) of this section.

(ii) The hearing conservation program shall include at least an annual audiometric test for affected employees at no cost to such employees.

(iii) If no previous baseline audiogram exists, a baseline audiogram shall be taken within 90 days for each employee (A) who receives a daily noise dose of 0.5 or above; or (B) who is required to wear hearing protectors pursuant to paragraph (f) of this section.

(iv) Each employee's annual audiogram shall be examined to determine if any significant threshold shift in either ear has occurred relative to the baseline audiogram.

(v)(A) If a significant threshold shift is present, the employee shall be retested within one month.

(B) If the shift persists: (*1*) Employees not having hearing protectors shall be provided with them in accordance with paragraph (f) of this section;

(*2*) Employees already having hearing protectors shall be retrained and reinstructed in the use of hearing protectors.

(*3*) The employee shall be notified of the shift in hearing level.

(2) *Audiometric testing*. (i) Audiometric tests shall be administered by a certified audiometric technician or an individual with equivalent training and experience.

(ii) Audiometric tests shall be preceded by a period of at least fourteen hours during which there is no exposure to workplace sound levels in excess of 80 dBA. This requirement may be met by wearing hearing protectors which reduce the employee noise exposure level to below 80 dBA.

(iii) Audiometric tests shall be pure tone, air conduction, hearing threshold examinations, with test frequencies including as a minimum, 500, 1000, 2000, 3000, 4000 and 6000 Hz and shall be taken separately for each car.

(iv) The functional operation of the audiometer shall be checked prior to each period of use to ensure that it is in proper operating order.

(v) Equipment, calibration and facilities shall meet the specifications set forth in the Appendix.

(h) *Information and warnings.*—(1) *Signs.* Clearly worded signs shall be posted at entrances to, or on the periphery of, areas where employees may be exposed to noise levels in excess of the limits prescribed in paragraph (c) of this section. These signs shall describe the hazards involved and required protective actions.

(2) *Notification.* Each employee exposed to noise levels which exceed the limits prescribed in paragraph (c) of this section shall be notified in writing of such excessive exposure within 5 days of the time the employer discovers such exposure. Such notification shall inform the affected employee of the corrective action being taken.

(i) *Records.* — (1) *Noise exposure measurements.* (i) The employer shall keep an accurate record of all noise exposure measurements made pursuant to paragraph (d) of this section.

(ii) The record shall include the following information: (A) name of employee, social security number and daily noise dose;

(B) location, date, and time of measurement and levels obtained;

(C) Name of person making measurement;

(D) Type, model and date of calibration of measuring equipment.

(iii) These records shall be maintained for a period of at least five years.

(2) *Audiometric tests.* (i) The employer shall keep an accurate record of all employee audiograms taken pursuant to paragraph (g) of this section.

(ii) The record shall include the following information: (A) Name of employee and social security number;

(B) Job location of employee;

(C) Date of the audiogram;

(D) The examiner's name and certification;

(E) Model, make and serial number of the audiometer; and

(F) Date of the last calibration of the audiometric test equipment.

(iii) These records shall be maintained for the duration of the affected employee's employment plus 5 years.

(3) *Calibration of audiometers.* (i) The employer shall keep an accurate record of all audiometer calibrations required to be made pursuant to paragraph (g) of this section and the Appendix.

(ii) The record shall include the following information:

(A) Type of calibration;

(B) Date performed; and

(C) All measurements obtained.

(iii) These records shall be maintained for a period of 5 years.

(4) *Access to records.* (i) All records required to be maintained by this section shall be made available upon request to authorized representatives of the Assistant Secretary and the Director.

(ii) Records of noise exposure measurements required to be maintained by this section shall be made available to employees and former employees and their designated representatives.

(iii) Employee audiometric data required to be maintained by this section shall be made available upon written request to the employee or former employee.

(j) *References.* (1) ANSI S1.4–1971, American National Standard Specification for Sound Level Meters, S1.4–1971, American National Standards Institute, 1430 Broadway, New York, New York 10018.

(2) ANSI S3.6–1969—American National Standard Specifications for Audiometers, S3.6–1969—American National Standards Institute, 1430 Broadway, New York, New York 10018.

(3) ANSI S1.11–1971—American National Standard Specification for Octave, Half-Octave, and Third Octave Band Filter Sets S1.11–1966 (Reaffirmed 1971), American National Standards Institute, 1430 Broadway, New York, New York 10018.

(4) ANSI Z24.22–1957 (R 1971)—American National Standard Method for the Measurement of the Real-Ear Attenuation of Ear Protectors at Threshold, Z24.22–1957, American National Standards Institute, 1430 Broadway, New York, New York 10018.

(5) American Industrial Hygiene Association Journal, 27:303–304 (May–June 1966). American Industrial Hygiene Association, 66 S. Miller Road, Akron, Ohio 44313.

(6) Council for Accreditation in Occupational Hearing Conservation, 1619 Chestnut Avenue, Haddon Heights, New Jersey 08035.

(7) ISO R389–1964—International Organization for Standardization Recommendation R389–1964, Standard Reference Zero for the Calibration of Pure Tone Audiometers, including Addendum 1–1970. Available from the American National Standards Institute, 1430 Broadway, New York, New York 10018.

APPENDIX

AUDIOMETRIC EQUIPMENT AND FACILITIES

(1) *Audiometric test rooms.* Rooms used for audiometric testing shall not have sound pressure levels exceeding those in Table G–16b when measured by equipment conforming to the requirements of ANSI S1.4–1971, Type 1 or Type 2, and ANSI S1.11–1971.

TABLE G–16b—MAXIMUM ALLOWABLE SOUND PRESSURE LEVELS FOR AUDIOMETER ROOMS

Octave band center frequency (Hz)	500	1000	2000	4000	8000
Sound pressure level (dB)	40	40	47	52	62

(2) *Audiometric measuring instruments.* (i) Instruments used for measurements required in paragraph (g) of this section shall be of the discrete frequency type which meet the requirements for limited range pure tone audiometers prescribed in ANSI S3.6–1969.

(ii) In the event that pulsed tone audiometers are used, they shall have a tone on-time of at least 200 milliseconds.

(iii) Self-recording audiometers shall comply with the following requirements: (A) The chart upon which the audiogram is traced shall have lines at positions corresponding to all multiples of 10 dB hearing level within the intensity range spanned by the audiometer. The lines shall be equally spaced and shall be separated by at least ¼ inch. Additional gradations are optional. The audiogram pen tracings shall not exceed 2 dB in width.

(B) It shall be possible to set the stylus manually at the 10 dB gradation lines for calibration purposes.

(C) The slewing rate for the audiometer attenuator shall not be more than 6 dB/sec except that an initial slewing rate greater than 6 dB/sec is permitted at the beginning of each new test frequency, but only until the second subject response.

(D) The audiometer shall remain at each required test frequency for 30 seconds (±3 seconds). The audiogram shall be clearly marked at each change of frequency and the actual frequency change of the audiometer shall not deviate from the frequency boundaries marked on the audiogram by more than ±3 seconds.

(E) For audiograms taken with a self-recording audiometer, it must be possible at each test frequency to place a horizontal line segment parallel to the time axis on the audiogram, such that the audiometric tracing crosses the line segment at least six times at that test frequency. At each test frequency the threshold shall be the average of the midpoints of the tracing excursions.

(3) *Audiometer calibrations.* (i) A biological calibration shall be made at least once each month and shall consist of testing a person having a known stable audiometric curve that does not exceed 25 dB hearing level at any frequency between 500 and 6,000 Hz and comparing the test results with the subject's known baseline audiogram, and

(ii) If the results of a biological calibration indicate hearing-level differences greater than 5 dB at any frequency, if the signal is distorted, or there are attenuator or tone switch transients, then the audiometer shall be subjected to a periodic calibration.

(iii) A periodic calibration shall be performed at least annually. The accuracy of the calibrating equipment shall be sufficient to assure that the audiometer is within the tolerances permitted by ANSI S3.6–1969. The following measurements shall be performed: (A) With the audiometer set at 70 dB hearing threshold level, measure the sound pressure levels of test tones using a National Bureau of Standards Type 9A coupler, for both earphones and at all test frequencies.

(B) At 1000 Hz, for both earphones, measure the earphone decibel levels of the audiometer for 10 dB graduations in the range 70 to 10 dB hearing threshold level. This measurement may be made acoustically with a National Bureau of Standards Type 9A coupler or electrically at the earphone terminals.

(C) Measure the test tone frequencies between 500 and 6000 Hz with the audiometer set at 80 dB hearing threshold level, for one earphone.

(D) A careful listening test, more extensive than that required for biological calibration shall be made in order to ensure that the audiometer displays no evidence of distortion, unwanted sound, or other technical problems.

(E) The functional operation of the audiometer shall be checked to ensure that it is in proper operating order.

(iv) An exhaustive calibration shall be performed at least every five years. This shall include testing at all settings for both earphones. The test results shall demonstrate that the audiometer meets specific requirements stated in the applicable sections of ANSI S3.6–1969 as listed below: (A) [Sections 4.1.2 and 4.1.4.3] Accuracy of decibel level settings of all test tones,

(B) [Section 4.1.2] Accuracy of test tone frequencies,

(C) [Section 4.1.3] Harmonic distortion of test tones,

(D) [Section 4.5] Tone-envelope characteristics, i.e., rise and decay times, overshoot, "Off" level,

(E) [Section 4.4.2] Sound from second earphone,

(F) [Section 4.4.1] Sound from test earphone, and

(G) [Section 4.4.3] Other unwanted sound.

(Secs. 6, 8, Pub. L. 91–596, 84 Stat. 1593, 1599 (29 U.S.C. 655, 657), Secretary of Labor's Order No. 12–71 (36 FR 8754)).

Signed at Washington, D.C., this 18th day of October 1974.

JOHN STENDER,
Assistant Secretary of Labor.

[FR Doc.74–24705 Filed 10–23–74;8:45 am]

Appendix C

Appendix C

OSHA COMPLIANCE PLAN

INTRODUCTION

________________________________ has developed the noise abatement program described herein to comply with the Occupational Safety and Health Act (OSHA) as promulgated by the U.S. Department of Labor. This program will safeguard the hearing of employees exposed to noise levels exceeding the federal criteria until such time as the levels are reduced by engineering, administrative, etc. procedures, below those that would normally produce a loss of hearing.

PROGRAM OF COMPLIANCE

The compliance program will include the following phases:

- •Educational
- •Noise Surveys
- •Engineering and Administrative Controls
- •Issuing of Personal Protective Equipment
- •Audiometric Testing

EDUCATIONAL PROGRAM

________________________________ intends to educate properly its existing work force and all new employees concerning the need for an effective hearing conservation program. The educational program will consist of an initial educational presentation by company personnel or qualified consultant on the need for an effective hearing conservation program, and an explanation of company policy relative to the requirements of wearing protective devices, etc.

It is anticipated that the educational phase will begin

on or before ______________________________.

SURVEYS

It is the intent of this company to conduct a detailed noise survey using instruments meeting the ANSI appropriate standard, and calibrated externally before and after each survey. The initial survey will include the measurement of sound levels using the appropriate scale and meter response and an octave band analysis at appropriate locations. The measurements will be made at the employees' normal working positions, and, in addition, a general survey will be made of the production area(s). This procedure will allow an approximate estimation of the employees' overall exposure in instances where the employee is required to move from one working position to another in his daily work routine.

Follow-up general surveys will be made yearly and at such a time as engineering, production, administrative, etc. procedures result in a change in the noise level of the environment. The results of all surveys will be properly recorded and made available for evaluation upon request.

It is estimated that the initial general survey will be completed on or before ______________________________.

ENGINEERING AND ADMINISTRATIVE CONTROLS

The possibility of controlling the existing levels through engineering control procedures will be properly considered within the educational and economic constraints of our organization. Due to the complexity of some of the equipment used by this company, it is realistic to assume that part of the burden for reducing the noise will rest upon the shoulders of the supplier. Therefore, suppliers of equipment exceeding the accepted guidelines normally purchased by this company will be notified directly or indirectly through the appropriate association(s) of which this company is a member, of the existing harmful noise conditions. The supplier will be requested to redesign his equipment where possible to meet the defined guidelines. As an interim solution, this company will consider possible redesign of existing equipment, the building of partial or total enclosures, and other engineering noise control procedures for reducing the existing noise levels.

Within the limitation of work schedules, employee training background, etc., possible administrative controls will be considered. Where practical, over-exposed employees will be moved at appropriate intervals to other work areas having noise levels significantly below the accepted guidelines. In addition, when practical, equipment operational procedures will be modified so that during any single production run, the allowed exposure will not exceed the allowed limit.

It is anticipated that engineering and administrative controls will begin to be implemented on or before

______________________________________.

ISSUING OF PERSONAL PROTECTIVE EQUIPMENT

Until such time as engineering and/or administrative controls reduce the amount of noise exposure below the allowed limits, appropriate personal protective equipment will be properly issued and made available. It is recognized that this equipment is considered to be a temporary solution to the problem and eventually will be eliminated.

As is required of all other safety equipment, the wearing of the ear protection provided will be required. Failure of the employee to wear the protection provided could result in the termination of his employment with the company.

It is anticipated that the issuing of ear protection devices will begin on or before ____________________.

AUDIOMETRIC TESTING PROGRAM

The objective of the hearing conservation program developed by ____________________________________, as stated previously, is the preservation of the hearing of its employees. In order to achieve this goal, an effective audiometric testing program will be initiated. It is anticipated that this program will include pre-employment audiograms (at least within ninety days of initial employment, during which time the employee will be provided and required to wear ear protection), an initial survey of the work force in order to establish a baseline, and termination audiograms when possible. All employees exposed to levels equal to or exceeding the published maximum noise level will be given an audiogram once a year.

It is anticipated that the initial testing of the existing work force will begin on or before ____________ __________________________.

Note: This Compliance Plan adapted from "A Recommended Noise Compliance Plan for Evaluation by Industry in Developing an Effective Hearing Conservation Program," W. C. Creel and L. H. Royster, North Carolina State University.

INDEX